THE OPEN UNIVERSITY

Science: A Third Level Course

Physiology of Cells and Organisms

Unit 5
Exocrine Secretion

Unit 6
Muscles and Contractility

Unit 7
The Cell Cycle

Prepared by an Open University Course Team

THE OPEN UNIVERSITY PRESS

S321

Course Team

Chairman and General Editor
Richard Holmes

Unit Authors
Vic Daniels
Gerald Elliott
Richard Holmes
Steven Rose
Jeff Thomas

Editor
Jacqueline Stewart

Other Members
Neil Chalmers
Norman Cohen
Vic Finlayson (*Staff Tutor*)
John Groom (*BBC*)
Stephen Hurry
Roger Jones (*BBC*)
Pat O'Callaghan (*Course Assistant, Day Schools*)
Barbara Pearce (*Course Assistant*)
Dave Spears (*Technician*)
Jim Stevenson (*BBC*)

Consultant
Bruce Carter, Department of Zoology,
University of Edinburgh

The Open University Press,
Walton Hall, Milton Keynes.

First published 1974.

Designed by the Media Development Group of the Open University.

Printed in Great Britain by
Martin Cadbury, a specialized division of Santype International, Worcester and London.

ISBN 0 335 02461 0

This text forms part of an Open University course. The complete list of units in the course appears at the end of this text.

For general availability of supporting material referred to in this text please write to the Director of Marketing, The Open University, P.O. Box 81, Walton Hall, Milton Keynes, MK7 6AT.

Further information on Open University courses may be obtained from the Admissions Office, The Open University, P.O. Box 48, Walton Hall, Milton Keynes, MK7 6AB.

1.1

Unit 5

Exocrine Secretion

Contents

Table A: part 1

Principal Sections of recommended prerequisite Courses on which this Unit depends

Course	Unit and Section No.	Topic
S100*	18	cells and organisms
S22–**	4.8	coordination of digestion
	4, Home experiment	histology of animal tissues
	8, TV programme	pheromones and autoradiography
	9.2.1	maintenance of blood sugar
	9.2.5	blood sugar and lactation
	9, Appendix 1	hormones and lactation

Table A: part 2

List of scientific terms, concepts and principles used in Unit 5

Developed in this Unit or its set reading	Page No.	Developed in a later Unit	Unit No.
acinus	9		
antibody	30	antibody	8
antigen	30	antigen	8
bicarbonate ion secretion	24		
casein	29		
centro-acinar cells	10		
chylomicron uptake by endothelium	31		
cisternal package (pathway)	21		
condensing vacuoles	12		
deoxycholate	11		
exportable protein	13		
fat droplet secretion by alveolar cells	33		
immunoglobulin	29	immunoglobulin	8 and 10
intralobular and interlobular ducts	9		
α-lactalbumin	34		
lactose formation in the Golgi	34		
perfused pancreatic duct	25		
plasma cells	30	plasma cells	8
pulse–chase experiment	15		
zymogen granules	10		

* The Open University (1971) S100 *Science: A Foundation Course*, The Open University Press.

** The Open University (1972) S22– *Comparative Physiology*, The Open University Press.

Objectives for Unit 5

After completing this Unit and watching the TV programme associated with it, you should be able to:

1 Define, recognize the best definitions of, or place in the correct context all the terms listed in Table A.

2 Describe in general terms the anatomy, histology and cell ultrastructure of the exocrine pancreas by labelling diagrams provided.

3 Outline the evidence of the nature of the intracellular secretory pathway of the pancreatic enzymes, which Palade and Siekevitz obtained by assaying the proteolytic activity of cell fractions in the guinea-pig.
(SAQ 1)

4 Describe (in 150 words) at least two pieces of evidence from work on the rabbit that challenge the view that the cisternal pathway is the only one in pancreatic enzyme secretion.
(SAQs 2 and 3)

5 Describe (in 250 words) two ways in which accurate information on the intracellular pathway and the timing of transport of pancreatic enzymes was obtained by the use of radioactive tracers.
(SAQs 4 and 5)

6 Describe (in 100 words) the evidence that the 'condensation' of zymogen granules from intracellular vesicles does not require the expenditure of respiratory energy or the synthesis of new protein.
(ITQs 3 and 4)

7 (a) Describe (in 150 words) how a primary secretion of bicarbonate and water might be produced by the acini of the pancreas.
(ITQ 5; SAQ 6)

(b) Describe (in 100 words) one piece of evidence indicating that the ducts may modify this secretion.
(SAQ 6)

(c) Describe (in 250 words) one piece of evidence that the water and bicarbonate components of the pancreatic juice may be produced *not* as a primary secretion in the acini but as a result of secretion by the ducts.

8 Show a knowledge of the anatomy, histology and fine structure of the secretory parts of the mammary gland by labelling diagrams provided.

9 Compare (in 500–1 000 words) the secretion of protein by the mammary alveolar cells and the pancreatic acinar cell, including at least two pieces of evidence for the selective nature of immunoglobulin secretion.
(SAQ 7)

10 (a) Describe (in 200 words) the process of milk fat secretion by the alveolar cell.
(SAQ 8)

(b) Describe (in 100 words) the secretion of lactose by the alveolar cell.
(ITQ 7)

(c) Compare (in 500 words) what was said about the secretion of water and electrolytes in the pancreatic acinar cell and the mammary alveolar cell.
(ITQs 9 and 10; SAQ 9)

11 Give at least four examples of the part played by the Golgi body in secretion by various types of cell.

5.0   Introduction and study guide for Unit 5

This Unit should take you rather less time than is usual for a full Unit; there is no set reading, nor are there any black-page offprints. Furthermore, we feel that the material is conceptually more simple than the material in Units 3 and 4 or 6, for example. In fact, Unit 6 is longer than average, so you should try to make a start on it in the same week as you complete this Unit.

The largest Section in this Unit, Section 5.2, deals with the pathway of protein secretion within the pancreatic acinar cell. In this Section we trace the development of the techniques used and describe the build-up of some of the relevant information between 1956 and 1971. You should at least read through this Section (and also Sections 5.5.2, 5.5.3 and 5.5.4) before watching the TV programme associated with the Unit, if you are to make the best use of the television.

The conclusions which can be drawn from Section 5.2 about the pathway of the secreted enzymes are simple, and are summarized in Figure 13. However, after working through this Section you should retain more than just the conclusions (see Objectives 3–6); the nature of the evidence offered and the techniques by which it was obtained are also important.

Section 5.3 is a short account of two of the views on how the bicarbonate and water are secreted, and Section 5.4 considers the control of the production of both enzymes and fluids.

Section 5.5 deals with milk secretion, and is intended to furnish you with some interesting comparisons and contrasts between the functioning of the pancreas and the mammary gland. Whereas Section 5.2 lays emphasis on *how* the results were obtained (most of them *in vitro*) Section 5.5 does not, as the type of experimental procedure used in the mammary studies is dealt with briefly in the latter part of the TV programme. (You should remember that in S321, as in many other Open University Courses, the television programmes form an integral, and therefore assessable, part of the Course.)

The previous two Units have been devoted to a detailed study of the plasma (outer cell) membrane, and the manner in which substances in solution may be transported across this membrane. Most of the examples considered have been concerned with the movement of relatively small molecules into the cell, and the processes involved probably occur in nearly all animal cells, though the intestinal cells mentioned may have an especially enhanced ability to take up some substances from the gut (see also Units 10 and 11).

There is no reason to suppose that the transport of substances out of the cell into the surrounding medium is a fundamentally different process—indeed the 'export' of, for example, sodium (Na^+) from the cell has been discussed at length. However, from previous Courses[1] you may recall that metazoans have cells (or whole organs) specialized for the production and release of substances into or out of the body on a comparatively massive scale—these are glands.

Gland cells are clearly very interesting in terms of the cellular transport processes discussed in Units 3 and 4, partly because transport across their plasma membranes must take place very rapidly and involve far greater amounts of substrate than would normally be the case in cells not specialized for secretion. In addition, the substrate transported may be a very large molecule, and the cell may be the site of the production of large quantities of it, as well as responsible for its transport across the plasma membrane either into the blood or on to some surface epithelium.

QUESTION   Can you recall from previous Courses one example of secretion of (a) a macromolecular substance into the blood, and (b) a macromolecular substance on to an epithelium?

ANSWER   Examples of (a) would be provided by many hormones, particularly protein ones—for instance, the secretion of gonadotrophins from the pituitary— and examples of (b) by many exocrine secretions—for instance, mucus from goblet cells into the gut, digestive enzymes from the pancreas into the gut.

What we are looking at in this Unit is, in a sense, the applied biology of the processes discussed in Units 3 and 4; we are considering the transport of small molecules, but in cells specialized as secreting 'factories'; we are considering 'bulk transport' (see Unit 4, Section 4.2) but here it involves the mass *export* of macromolecules; finally, we are considering the physiological significance and control of these transport phenomena, that is, how they are coordinated to meet the needs of the whole animal. You might have gained the impression from the elegant work cited in the preceding Units that even in animals such as the mammals, the cells of the tissues are leading independent, well regulated, lives, quite separate from one another.

This is a fair view—up to a point. Given that a constant environment bathes the cells, and assuming that no stimulus produces a major change in their metabolism, cells will regulate their own internal environments, respire and repair themselves much as independent unicells might. However, the constancy of the fluids bathing the cells is only maintained by homeostatic mechanisms which involve coordinated changes in the metabolism of groups of cells; what is more, these coordinated changes more often than not are in fact partly changes in the membrane transport systems of the cells. One of the examples of this, with which you are familiar, is the increased uptake of blood glucose and fatty acids by muscle cells in response to the action of insulin.[2] Thus, the uptake of glucose may not always be on an independent cell-by-cell basis, but may be regulated by the organism as a whole.

In this same example, perhaps the more striking instance of 'organism-regulated' transport is provided by the release of the insulin itself, from the β-cells of the pancreas. The insulin, a protein, is transported from within the β-cell to the blood after changes in the glucose level of the blood circulating through the pancreas.

5.1 Selection of the tissues to serve as examples of secretion

You are already familiar with a number of secretory tissues, both endocrine and exocrine. Any of these should serve to illustrate the processes mentioned above— but Unit writing resembles politics in being the 'art of the possible' and thus being a compromise with the ideal situation. The fact is that secretion has been examined extensively in only a few tissues; furthermore, it is too early to be able to say that the secretion of similar substances by different organs necessarily involves precisely the same processes.

For these reasons we have chosen to consider secretion in two exocrine glands, the exocrine pancreas and the mammary gland. They are glands you already know a little about;[3,4] both secrete proteins as well as water and electrolytes, yet they are controlled differently and have very different functions. However, even though we have, so to speak, hand-picked the glands, you will find that there are large areas of their functioning which are either still unknown or else largely a matter for speculation. We hope this doesn't disturb you; if it does, perhaps you should be doing physics!

5.2 The exocrine pancreas

From a physiological viewpoint, the pancreas has two major functional secretory parts. The majority of the tissue forms the *exocrine gland*, whose secretion is collected into the pancreatic duct and discharged into the small intestine.[4] Scattered about within the exocrine gland are the islets of Langerhans[2] which are made up of groups of *endocrine gland* cells, some (the β-cells) producing insulin and other (α-cells) producing glucagon. It is with the exocrine pancreas that we are concerned here.

The exocrine pancreas is interesting in that in some ways it functions rather like two separate glands with distinct control mechanisms. You may recall[4] that at least two hormones are involved in the regulation of pancreatic* secretion:

* Unless otherwise stated, for the rest of this Unit 'pancreas' should be taken to refer to the exocrine pancreas.

secretin and pancreozymin.* When purified extracts of these hormones are injected into fasting animals, secretin produces a copious flow of pancreatic fluid consisting of electrolytes (with a particularly high concentration of HCO_3^-) and water, whereas pancreozymin stimulates the production of a viscous fluid very rich in enzymes.

pancreatic fluid

A study of this gland might be broken down into three main areas:

1 How do cells within the gland synthesize and secrete protein enzymes?

2 How does the gland produce a large flow of water and bicarbonate ions?

3 How do the physiological control mechanisms actually act on the cells to regulate secretion and thus coordinate it with the animal's eating habits?

We will approach these three areas, in this order, in the following three Sections; however, it is not possible to separate them entirely, and aspects of 3 are likely to creep into any discussion of 1 and 2.

5.2.1 Structure of the gland

The structure of the exocrine pancreas comes close to being that of a 'classic' exocrine gland. Secretory cells are arranged in groups around a small, and not always obvious, lumen or central space (see Fig. 1). (You saw a rather similar arrangement in the Brunner's glands present in the small intestine.[5]) The groups of cells form spheres, or spheres that are elongated towards the cylindrical. Each of these groups is called an *acinus* (Latin for a grape) and the cells making up an acinus are referred to as *acinar* cells. The acini are connected by very delicate ducts, the *intralobular* ducts, and thus are held to resemble a microscopic bunch of grapes. This simile may help you to visualize the basic structure, as well as refute any suggestions that the Course Team is unaware of the existence of plants!

acinar cells

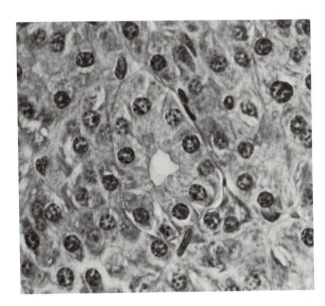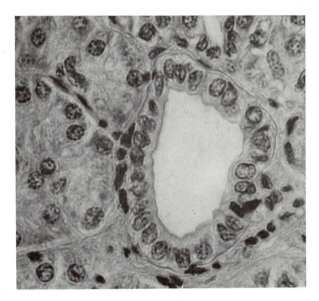

The 'bunch of grapes' as a whole forms a *lobule*, and these small lobules are connected by *interlobular ducts* (Fig. 2). The interlobular ducts join the pancreatic duct (Fig. 3), a major duct which runs right through the pancreas and joins the duodenum. There are, of course, some minor variations among species, even among mammals. For instance, there may be more than one 'pancreatic duct' entering the duodenum. The pancreas arises during embryonic development as one dorsal and one ventral outgrowth from the epithelium of the duodenum; these then extend and branch to form the acini and ducts. It is therefore not surprising that some species maintain both original connections with the gut.

The cells of the intralobular ducts may project into the lumen of the acinus,

Figure 1 (*left*) Photomicrograph of pancreatic acini. The nuclei of 11 acinar cells are visible, grouped around the lumen of a single acinus in the centre of the field. ($\times450$)

Figure 2 (*right*) Photomicrograph of an intralobular duct. Cuboid epithelium lines the duct, which has a diameter little larger than that of a single acinus. ($\times450$)

---

* Pancreozymin is now thought to be identical to the hormone cholecystokinin, which stimulates the release of bile from the gall bladder. Thus pancreozymin is often referred to in the literature as CCK-PZ. We have not adopted this convention.

filling much of it. These are called the *centro-acinar* cells, for obvious reasons. Their function is discussed in the next Section.

centro-acinar cells

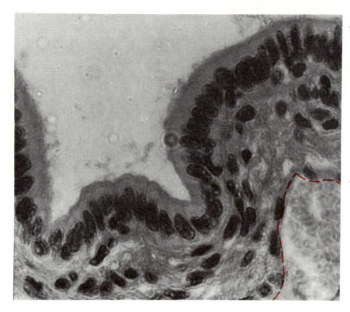

Wall of duct

Figure 3 Photomicrograph of part of the wall of a branch of the pancreatic duct. The duct is lined with columnar epithelium, and the nuclei of the underlying connective tissue are visible. (×450)

5.2.2 Secretion of protein by the exocrine pancreas

It has been known for a very long time that the pancreas produces a digestive juice which, among other things, attacks proteins. Nothing much was known of the manner in which the gland worked until 1875, when Heidenhain produced the hypothesis that precursors of the protein-splitting substances were produced within the acinar cells, and released after a meal was eaten. He based this on microscopical observations which showed that (a) there were numerous granules in the cytoplasm of the acinar cells of animals killed before they were fed, (b) these could not be seen in animals that had recently eaten and (c) that they reappeared in the cells of animals killed a few hours after a meal. By following the time-sequence of this disappearance and reappearance carefully, and by simultaneously assaying the proteolytic activity of the juice that the animal was producing at the time it was killed, Heidenhain concluded that the granules were undoubtedly precursors of the proteolytic substance, and he called them zymogen* granules.

zymogen granules

Not a great deal more was learned about enzyme secretion from the pancreas for the next 70 years. Then Heidenhain's hypothesis was confirmed when it was shown that substances such as pancreozymin which stimulated digestive secretion, also cleared the acinar cells of zymogen granules, and that if these granules were separated from the cells by techniques involving centrifugation, etc. (see Unit 1), and put in an appropriate medium, they showed hydrolytic activity. It was also shown that many enzyme precursors were present—including those of lipase, amylases, trypsin, chymotrypsin. Thus the source of the enzymes seemed clear.

It was shown in 1946 that, when administered *in vivo*, the chemical ethionine will destroy the acinar cells of the pancreas, but not the duct cells; it was also observed that after treatment with ethionine, enzyme secretion was severely reduced, but HCO_3^- and water secretion were not.

It had also been shown that if the pancreatic duct is tied off the acinar cells degenerate quite quickly, but the cells of the ducts and the centro-acinar cells do so much more slowly. If the duct is untied as soon as the acinar cells have degenerated, a flow of water and bicarbonate can still be produced, although enzymes cannot. This suggests that the centro-acinar cells or the duct system may be responsible for the fluid production.

The state of knowledge stayed much at this level until the advent of electron microscopy, which gave great impetus to the investigation of intra-cellular events.

* ZYMOGEN: a slightly dubious word meaning 'enzyme-ogenic' or 'enzyme-producing'.

The combined use of the electron microscope and biochemical techniques in this field was pioneered by George Palade, who first started work on pancreatic secretion in the early 1950s and who, with other distinguished co-workers, is probably still the leader in this field.

In 1956 Palade, working originally with Siekevitz, set out to establish where the contents of the zymogen granules were made, how they became zymogen granules and how they were secreted into the pancreatic duct. They were not alone in these endeavours, as other groups joined in and the field became an active one. Also, they were working at a most propitious time, when information and speculation on the location and method of protein synthesis were appearing at a great rate. (Indeed, within that general area of science the speed of transcription of genetic information was rivalled only by the rate of translation of biochemists into Nobel Laureates!)

Light microscopy had revealed that when the acinar cell is stained, the region around and below the nucleus is strongly basophilic. Electron microscopy and histochemical studies showed that this is due to the RNA on very densely packed rough endoplasmic reticulum (RER). Above the nucleus, the Golgi apparatus is strikingly well developed. This appearance—highly developed basal RER and Golgi apparatus—is in fact common in cells secreting proteinous substances.

Palade and Siekevitz found that there were differences between the appearance of the RER of the acinar cells in guinea-pigs which had been starved for 48 hours and those which had been killed an hour after feeding. In the former, the RER is tightly packed, tends to be orientated in a particular plane and to have contents which appear pale in electron micrographs. Within an hour of feeding 40–60 per cent of the acinar cells show changes in their RER. The 'preferred' orientation is lost, the cavities appear distended, and within them there appear many larger, dense, granules.

feeding and zymogen granules

QUESTION Bearing in mind Heidenhain's original observations on the state of the acinar cells before and after feeding, what do you consider is a likely explanation for these changes in the RER?

ANSWER After 48 hours of starvation, the apical parts of the acinar cell will be full of zymogen granules. It could well be, therefore, that production of new enzyme by the RER has slowed or stopped. In the fed animals, the zymogen granules will have been shed; the swelling of the cisternae (cavities) and the inclusion in them of dense granules certainly suggests that secretion of new, probably proteinous, material is taking place.

To attempt to verify whether these granules within the cisternae of the RER were in fact the precursors of the zymogen granules, Palade and his colleagues (1956) took the pancreas from guinea-pigs that had been starved for 48 hours and then fed 1 hour before the experiment, and prepared a homogenate; they also did the same with cells from animals that had not been fed. They then fractionated the homogenates, and found they got five distinct fractions. These were: a nuclear fraction (1), a zymogen fraction made up of tightly packed zymogen granules (2), a mitochondrial fraction (3), a microsomal fraction, made up of fragmented endoplasmic reticulum (4) and a final supernatant fraction, representing the cytoplasmic matrix (5).

fractionation of pancreatic cells

When they assayed the proteolytic activity of the various fractions (the enzymes are of course present as inactive precursors, e.g. trypsinogen, chymotrypsinogin, which have to be activated—in this case by treatment with active trypsin), they found that in the starved animals the activity of the zymogen fraction was high and that of the microsomal fraction very low. In the fed animals, the activity of these two fractions was closely similar. In both fed and unfed animals the activity of the cytoplasmic fraction (fraction 5) was low—not more than 13 per cent of the total proteolytic activity.

activity of the cell fractions

To isolate the active elements within the microsomal fraction more precisely, attempts were made to separate the intra-cisternal granules from the rest of fraction 4. The fraction was first treated with a low concentration (0.1 per cent) of deoxycholate—a detergent.

QUESTION What general action would you expect a detergent-like chemical to have on the microsomes?

ANSWER It would be reasonable to suppose that it would attack the lipid of the membranes, dissolving it and thus largely destroying the membranes —or in small amounts, perhaps just making them more permeable.

After this treatment, fraction 4 was subjected to differential centrifugation, which produced, in the homogenate derived from fed animals at least, three distinct sub-fractions: (i) a heavy one containing mainly granules, but also some broken membrane debris; (ii) a medium one consisting mainly of ribonuclear protein particles and (iii) a solution containing the dissolved parts of the membranes.

In the heavy sub-fraction (sub-fraction (i)) from fed animals, proteolytic activity was more concentrated than in the original complete microsomal fraction. In starved animals, the heavy sub-fraction was not only much smaller, but showed no concentration of activity when compared with the 'parent' microsomal fraction— which, as you will recall, itself showed much less activity than in fed animals.

The result certainly seemed to lend support to the idea that the enzyme precursors are made on the RER (presumably on the ribosomes) and transported through the cisternae of the endoplasmic reticulum (ER) to the apical regions of the cell, where they are 'condensed' to form the relatively large zymogen granules. Electron micrographs (see Fig. 4) showed that while the small granules were in

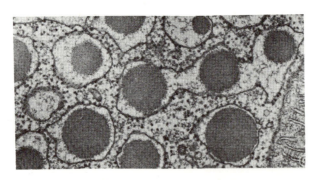

Figure 4 Electron micrograph of RER cisternae containing electron-dense granules of proteolytic enzyme. ($\times$ 50 000)

proposed intracellular pathway of secreted enzymes

the RER cisternae they appeared to be without a limiting membrane. Later, when they had formed larger granules further up the cell, they were surrounded by a membrane. These membrane-surrounded granules appeared rather less dense than the zymogen granules they finally became (or gave rise to); for this reason they were termed 'condensing vacuoles' as it appeared that the enzyme precursors were progressively concentrated within them. However, as the zymogen granules were also larger, it seemed possible that some fusion of condensing vacuoles might also be taking place.

This work clearly suggested a possible route for the secretion of protein— manufacture on the ribosomes, transport through the cisternae to form condensing vacuoles and then zymogen granules in the apical regions of the cell, followed by release of the granules into the acinar lumen after the arrival of the appropriate stimulus.

However, it left unanswered some very important questions:

1 How does the protein, presumably manufactured on the ribosomes on the cytoplasmic side of the RER membrane, get into the cisternae of the RER? Can you recall any discussion in the previous Units of a mechanism of transport of whole protein molecules across a membrane, that might apply here?

2 Is this route the only, or even the major, route for the export of protein into the acinar lumen?

3 How, precisely, do the small granules in the RER move to the apical regions of the cell, and how do they grow into condensing vacuoles?

4 How do the condensing granules become zymogen vacuoles?

5 How are the zymogen granules released from the cell?

In the following Section we look very briefly at some of the attempts to answer these questions.

5.2.3 Investigation of the pathway of protein secretion *in vitro*

(a) *RER → cisternae*

The simple answer to the first question posed above—how does the exportable protein get from the ribosomes into the cisternae—is that we don't yet know. Three fairly obvious models can by suggested. First, there might be a large 'fixed' pore, through which the large enzyme molecules could pass (see Fig. 5); secondly, there might be a pore which is 'guarded', that is, a fugacious pore, involving a molecule which 'recognized' the enzyme and allowed it through, effectively sealing the pore against other molecules (see Fig. 6) or thirdly, a piece of ER might become 'pinched-off' with the enzymes included (see Fig. 7).

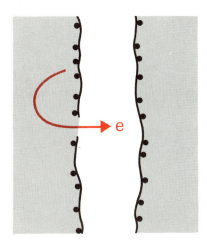 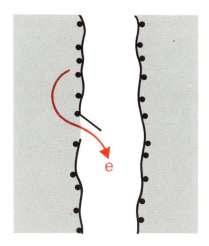 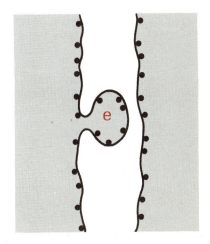

There are some difficulties associated with the acceptance of any of these models. Take the permanant pore—other molecules (e.g. non-exportable proteins) would be expected to move into the cisternae in significant amounts; but there is no evidence that this happens. With the fugacious pore, the mechanism is hard to visualize and the kinetics present problems. For example, unless the rate at which enzymes can be transported is always as great as (or greater than) the rate of its production on the ribosomes, it would be expected that at times of maximum manufacture there would be a 'back-up' of enzymes on the cytoplasmic side; but this has not yet been shown to happen. The third model is quite an attractive one, and something of the sort seems to occur in biological systems elsewhere (see Unit 10 on the uptake of protein by the gut in the new-born mammal). This model is simple to visualize as involving the features of pinocytosis discussed in Unit 4, Section 4.2. However, there are some strong arguments against it. For example, the actual process has not been seen occurring in any of the many appropriate electron micrographs which have been made. Furthermore, it would be expected that the 'pinched-off' membrane, presumably forming the apparent cisternal wall, would be 'rough' on the inside—that is, have ribosomes folded inwards—but this, too, has not been seen (Fig. 4).

Thus any answer to this first question must, at the moment, be highly speculative.

(b) *Is the cisternal pathway the only one for exportable protein?*

There is now no doubt that enzymes (or more strictly speaking, their precursors) are transported from the RER, in 'cisternal packages', to the zymogen granules, by routes we shall discuss later in this Section. These enzymes include trypsinogen, chymotrypsinogen(s) and ribonuclease.

What is less certain is whether all of the exportable fraction of these enzymes goes by this route, and whether all other exportable enzymes follow the same route. The work of Palade and others, quoted previously, on the secretion of proteolytic enzymes in the guinea-pig, clearly offers some relevant evidence on the first uncertainty.

ITQ 1 What is this evidence?

Clearly any technique for investigating the physiology of a structure by homogenizing it and then separating the pieces in a centrifuge is bound to be crude in some respects. The fraction representing the cytoplasmic matrix can only be

Figure 5 (*left*) Fixed pore model of transport of enzyme into the cisternae.

Figure 6 (*centre*) Fugacious pore model of transport of enzyme into the cisternae.

Figure 7 (*right*) The transport of enzyme by the extrusion of part of the RER into the cisternae.

13

said to consist *predominantly* of that matrix; it is highly probable that it will also contain material broken off the ER and other organelles because of the rough treatment it has received (see the TV programme for Unit 1). For this reason Palade and his co-workers, and many others, considered that this much activity in the supernatant was well within the experimental error of the technique and would be expected even if *all* the enzymes were either attached to ribosomes or cisternally packaged in the intact cell.

This is probably a fair assumption, and it can be held that their results provide quite good evidence that the proteolytic enzymes, notably trypsinogen (Tg), chymotrypsinogen (ChTg) and exportable ribonuclease are normally 'packaged', in the guinea-pig at least. It does, nevertheless, emphasize one of the problems of physiological research conducted *in vitro*.

Other workers have looked at the distribution of these enzymes in other species, using broadly similar techniques, but not always with exactly similar results. For example (Morris and Dickman, 1960), the cytoplasmic fraction of mouse pancreas is reported to contain a high proportion of the total exportable ribonuclease. Rothman (1969) found that homogenates of rabbit pancreas showed distributions of ChTg (only 17 per cent in the supernatant) similar to those found by Palade in the guinea-pig, but he found that there was 42 per cent of the total ChTg in the cytoplasmic fraction of the rat, using identical techniques. This raises the possibility of quite fundamental differences among species. (But it is possible that these differences are artefacts—the RER might lose proteins to the matrix more readily in the rat than in the rabbit and guinea-pig.)

Rothman reasoned that if all the enzymes in the cytoplasmic supernatant were derived from ruptured RER, etc., then they would be expected in the same *relative* proportions as in the microsomal fraction—or for that matter in the zymogen granules. He therefore measured (in the rabbit) the percentage of the total Tg that was found in the cytoplasm and compared it to the percentage of the total ChTg found in the cytoplasm. (He chose to compare Tg with ChTg rather than any other enzyme because the molecules are very similar, and thus the likelihood of a differential 'breaking away' from the ER was reduced.) As we have already mentioned, the percentage of the total ChTg which was found in the supernatant was 17 per cent, but some 32 per cent of the Tg appeared in the supernatant fraction.

Rothman also examined the effect on the relative proportions of the enzymes in the supernatant after the pancreas had been stimulated to very vigorous secretion by the injection of pancreozymin into the rabbit before it was killed. The ratio of Tg to ChTg in the particulate fractions remained the same (though, of course, their absolute values changed) but the ratio in the supernatant changed, and a relatively large amount of the trypsinogen was lost. Whereas in the control animals the ratio of the amount of Tg in the supernatant to the amount of ChTg in the same fraction was 6.5 : 1, in the pancreozymin-treated animals it had fallen to 4.6 : 1 (see also Table 1).

Table 1   Comparison of Tg and ChTg in fractionated rabbit pancreas

	Percentage of total recovered enzyme activity found in supernatant	
	Control	Pancreozymin-treated
Trypsinogen	32 ± 2 (34 ± 2)	(24 ± 3)
Chymotrypsinogen	17 ± 3 (20 ± 3)	(19 ± 3)
Ratio of activity of Tg/ChTg	(6.5 ± 0.6)	(4.6 ± 0.4)

N.B. The figures in brackets refer to four paired experiments of the control/pancreozymin group.

Bearing in mind the caveats we have raised about possible artefacts, these results do not constitute proof that an alternative, cytoplasmic, pathway exists for the secretion of, for example, trypsinogen. However, they do suggest the need to keep an open mind to this possibility and not to become quite carried away by the force and elegance of the 'cisternal package' model and assume that nothing else exists.

For some of the other enzymes found in the secretion of the gland, the situation is as uncertain. For example, Hokin (1955) found almost 85 per cent of the total exportable amylase from the pigeon pancreas in the cytoplasmic fraction.

(c)   *How do the small granules leave the RER and become condensing vacuoles?*

One of the later pieces of evidence supporting the (1956) proposition that the exportable protein was made on the ribosomes of the RER and then stored in the zymogen granules was provided by Siekevitz and Palade (1960). They showed that if they injected a fed guinea-pig with the appropriate amino acids labelled with a radioactive isotope, the label could be found (by fractionation) initially in the proteins associated with the RER and subsequently in the zymogen granules.

To try to define the intermediate route more precisely, Caro and Palade (1964) approached the problem with electron microscopic autoradiographic techniques. They injected the ^{14}C-labelled amino acid into the intact animal, killed it, and made autoradiographs of the exocrine cells. The autoradiographic grains were found in the ribosomes, in the RER and apparently in the peripheral vesicles of the Golgi complex, as well as in the larger condensing vacuoles and zymogen granules. This evidence seemed to implicate the Golgi complex as the link between the RER and the large zymogen granules, and indeed suggested that the condensed vacuoles might be simply a rather large variety of Golgi vacuole (Unit 1, Section 1.4.5).

EM and autoradiography

Only a limited amount of information could be gained from this approach, however, and one of these limits was imposed by the resolution available from autoradiographic techniques—it was difficult to make firm conclusions when structures smaller than 0.2μm were involved. Furthermore, few definite conclusions could be drawn about the kinetics of the process; for adequate 'label' to be taken up by the acinar cells, it had to be circulating in the animal for at least 15 minutes before the animal was killed. Thus some of the radioactive granules found in the cell would have been there for 15 minutes, others only seconds, but there was no way to distinguish them. Nor was it possible to ascertain whether changes in the location of radioactive granules within the cell were due purely to intracellular transport or partly to differences in the sites of synthesis.

ITQ 2   Can you think of some modification of the techniques used by Caro and Palade, and Palade and Siekevitz, that would enable a more precise analysis of the intracellular steps to be made?

Jamieson and Palade (1967a) combined fractionation techniques with EM auto-radiography and a system in which they incubated thin slices of guinea-pig pancreas in an appropriate life-support medium.* From examination of samples of the thin slices by electron microscopy, they concluded that the great majority of the acinar cells in the slices remained intact for at least 3 hours in the medium and, so far as could be seen, structurally normal. The general procedure they adopted was to incubate the slices for 3 minutes in a medium, called the 'pulse' medium in which one of the amino acids was labelled (L-leucine ^{14}C). The slices were then rinsed and put in the 'chase' medium, which was a medium basically identical to the pulse medium, but in which the only leucine was of unlabelled (i.e. ^{12}C) L-leucine. The slices were then incubated in the chase medium for various periods of time (e.g. 7, 17 and 57 minutes) and then either sectioned and photographed in the EM, or homogenized, and fractionated. Some microsomal fractions were themselves sectioned and examined under the EM and the specific radioactivity of the protein in corresponding samples was also measured.

sliced pancreas *in vitro*

pulse–chase

The fractionation technique used is summarized in Figure 8, and involved refinements which enabled Jamieson and Palade to separate rough microsomes (i.e. RER cisternae) from smooth microsomes with a higher degree of purity. Even so, the 'total microsome' fraction separated into three bands, the middle one of which (Band II) contained a mixture of particles; they therefore used a system of selective sampling making use of Band I (smooth microsomes) and Band III (rough microsomes) even though this meant that they were taking representative samples only.

---

* The incubation medium was a Krebs–Ringer bicarbonate solution equilibriated in 95 per cent O_2, 5 per cent CO_2, with 14 mM glucose and a pH of 7.6, with added L-amino acids.

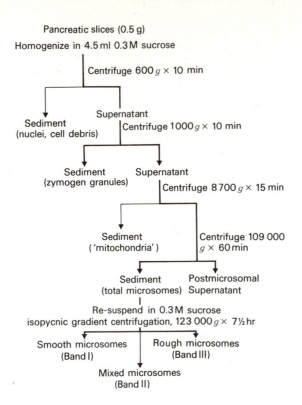

Pancreatic slices (0.5 g)
Homogenize in 4.5 ml 0.3 M sucrose

Centrifuge 600 *g* × 10 min

Sediment
(nuclei, cell debris)

Supernatant
Centrifuge 1000 *g* × 10 min

Sediment
(zymogen granules)

Supernatant
Centrifuge 8700 *g* × 15 min

Sediment
('mitochondria')

Centrifuge 109 000
g × 60 min

Sediment
(total microsomes)

Postmicrosomal
Supernatant

Re-suspend in 0.3 M sucrose
isopycnic gradient centrifugation, 123 000 *g* × 7½ hr

Smooth microsomes
(Band I)

Rough microsomes
(Band III)

Mixed microsomes
(Band II)

Figure 8 The cell fractionation scheme of Jamieson and Palade.

As the Golgi complex and the vesicles apparently produced by it make up far and away the largest proportion of smooth endoplasmic reticulum (SER) and smooth vesicles in these cells, Jamieson and Palade assumed that the smooth microsomal fraction could be considered essentially as Golgi vesicles and fragments of Golgi cisternae.

Using these methods, they were able to follow the short pulse of radioactive leucine (and the protein into which it was incorporated) through the cell with some degree of accuracy.

Table 2 Specific radioactivities (measured as counts per minute/mg) of proteins in microsomal and submicrosomal fractions and in the post-microsomal supernatant

Exp. No.	Incubation time in minutes		Total microsomes cpm/mg protein	Smooth microsomes cpm/mg protein	Rough microsomes cpm/mg protein	Post-microsomal supernate cpm/mg protein
	Pulse	Chase				
1	3	0	2 750	2 220	5 080	—
	3	+ 7	2 090	5 100	2 480	—
	3	+17	1 450	2 580	1 850	—
2	3	0	2 180	1 680	2 740	360
	3	+ 7	2 020	3 500	1 670	470
	3	+17	1 365	1 850	1 350	440
3	3	0	3 380	1 720	3 950	390
	3	+17	3 510	3 770	2 050	650
4	3	0	3 670	2 390	5 370	213
	3	+17	3 200	5 780	2 590	450
5	3	0	3 050	980	4 540	300
	3	+17	1 950	3 480	2 030	480
6	3	0	1 860	470	2 310	210
	3	+17	2 140	1 612	2 420	280
7	3	0	1 370	920	2 300	370
	3	+17	1 070	2 000	1 120	460
8	3	0	3 400	1 460	3 800	990
	3	+17	2 530	3 770	2 300	970
	3	+57	1 750	2 190	1 470	930

movement of radioactive label through the cell

16

Table 2, which is a shortened version of one appearing in Jameison and Palade (1967a), summarizes the sequence of transfer of labelled protein from the cisternae of the RER to the Golgi complex. Note that the values represent the specific activity of the samples, that is, the activity per mg of protein, not the total activity of the fraction.

From Table 2 it appears that in fed animals the newly manufactured enzyme begins to appear in the Golgi complex within a very short time (some time during the 3-minute pulse and rinsing time) but the main transfer occurs within 7 minutes after the end of the pulse.

It is not clear exactly how this transfer is brought about. It has, however, been observed that in a number of cells, including the acinar cells of the pancreas, there is a sort of 'transitional' ER around the peripheral region of the Golgi complex. The ER in this region is partly smooth and partly rough and is in close proximity to the small smooth-membraned vesicles forming the (outer) surface of the Golgi complex. It has been suggested that granules may be carried into the smooth areas of the ER, which then 'buds' off (such 'buds' or 'blebs' have been described; see Fig. 9, A and A_1) to form the smooth vesicles containing the enzymic granules. It is known that this step, RER → Golgi vesicles, requires energy. Alternatively, there may be what are in effect tubules, which are open permanently or periodically, between the 'transitional' area of the ER and the Golgi complex; some authors believe they have seen such structures in other secretory cells, but there is no evidence for them in this area of the guinea-pig pancreatic acinar cell.

budding of SER into Golgi complex

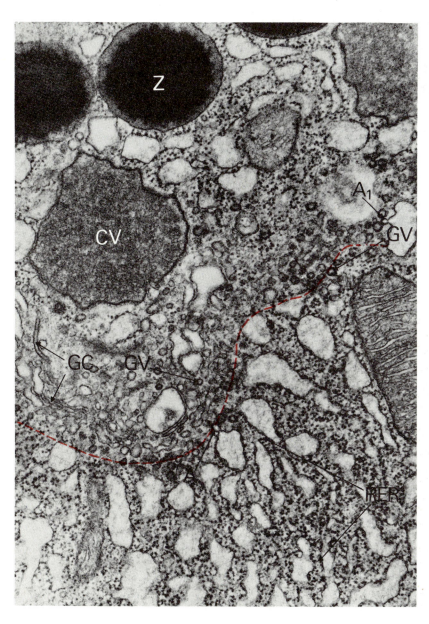

Figure 9 Electron micrograph showing a region of the cell apical to the nucleus. GC—Golgi cisternae, GV—Golgi vesicles, Z—zymogen granule, CV—condensing vacuole, A and A_1 —apparent 'buds' from RER to form Golgi vesicles. ($\times 50\,000$)

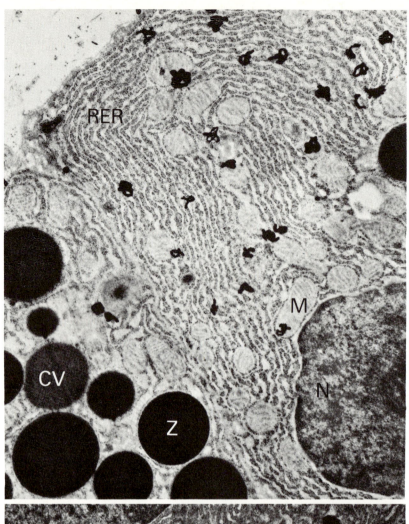

Figure 10 (×17 000)

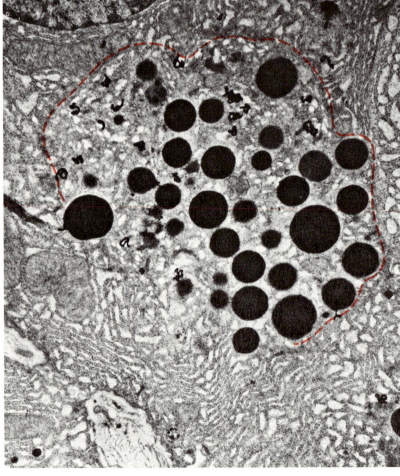

Figure 11 (×17 000)

18

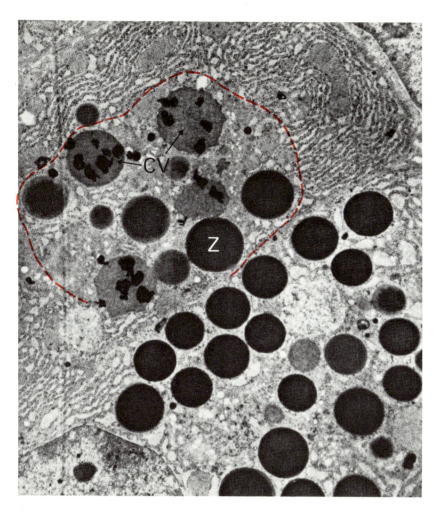

Figure 12 (/ 13 000)

Figures 10, 11 and 12 show electron microscope autoradiographs of thin slices of pancreas obtained by Jamieson and Palade. Figure 10 shows the distribution of labelled leucine at 0 minutes after the incorporation of the pulse (the label is confined to the RER). In Figure 11, +7 minutes after incorporation, the labelled leucine arrives at the Golgi complex. At +37 minutes the label appears in the condensing vacuoles (CVs) (Fig. 12) Note: the red dotted line indicates the periphery of the Golgi complex.

A point which seems to distinguish enzyme transport in the guinea-pig pancreas from the secretory processes in some other cells is that no trace of the enzyme has been found in the actual stacked cisternae of the Golgi apparatus. It seems that it is only the peripheral vesicles of the Golgi which are involved at this stage. This point is discussed further in the TV programme associated with this Unit.

absence of enzyme in Golgi cisternae

The process by which the peripheral vesicles become condensing vacuoles, is at the moment obscure; the larger size, and the location within the Golgi complex of the condensing vacuoles (Fig. 9) would suggest that perhaps numbers of vesicles fuse together.

Jamieson and Palade's work (1967b) has, however, given us a clear indication of the time course. As judged by EM autoradiographs made on tissue slices, (Figs. 10, 11, and 12) a single 3-minute pulse of radioactive leucine was arriving at the Golgi complex at +7 minutes and peaking (57 per cent of the visible grains) around +17 minutes (see also Table 3); between +17 and +37 minutes a significant proportion had left the peripheral vesicles and had been incorporated in recognizable condensing vacuoles, though not yet in the zymogen granules. From +37 to +57 minutes about 49 per cent of the visible grains had moved on to the condensing vacuoles, and label had begun to appear in the zymogen granules nearest the Golgi complex. Between +57 minutes and +117 minutes, the great majority of the grains were to be found on zymogen granules, and by +117 minutes the grains were heavily concentrated on the zymogen granules near the apex of the cell, and also in the acinar lumen and the collecting ducts of the gland. The Golgi complex was virtually free of autoradiographic grains (see Table 3 overleaf).

Table 3 Distribution of autoradiographic grains over cell components

	3-min (pulse)	percentage of autoradiographic grains				
		Chase incubation				
		+7 min	+17 min	+37 min	+57 min	+117 min
Rough ER	**86.3**	43.7	37.6	24.3	16.0	20.0
Golgi complex*						
Peripheral vesicles	2.7	**43.0**	37.5	14.9	11.0	3.6
Condensing vacuoles	1.0	3.8	19.5	**48.5**	35.8	7.5
Zymogen granules	3.0	4.6	3.1	11.3	32.9	**58.6**
Acinar lumen	0	0	0	0	2.9	7.1
Mitochondria	4.0	3.1	1.0	0.9	1.2	1.8
Nuclei	3.0	1.7	1.2	0.2	0	1.4
No. of grains counted	300	1 146	587	577	960	1 140

The boldfaced numbers indicate maximum accumulation of grains over the corresponding cell component.

* At no time were significant numbers of grains found in association with the flattened, piled cisternae of the complex.

(d) *How do the condensing vacuoles (CVs) become zymogen granules?*

Viewed under the electron microscope, zymogen granules appear more electron-dense and more regular than the condensing vacules. Autoradiographic pictures confirm that there is a progressive increase in the concentration of enzyme in the sequence young CV → older CV → zymogen granules. They are surrounded by a unit membrane closely resembling that of the plasma membrane and the evidence available suggests that, in the guinea-pig, enzyme secretion takes place by the release of the contents of zymogen granules rather than by the release of condensing vacuoles or Golgi vesicles. (It is not certain that in rapid continuous secretion induced by artificial stimulation, normal zymogen granules always appear, and as we mentioned earlier, it has not been shown beyond doubt that no enzyme is secreted via the cytoplasm.) There are probably some real functional differences between condensing vacuoles and zymogen granules—the distinction is more than just the pedantry displayed by some micro-anatomists. If this is so, it may be of some importance to know how the enzymic material is condensed and the outline achieved.

Using the techniques already described in this Section, Jamieson and Palade (1971) investigated the effect of certain inhibitors on the 'condensing process'— the conversion into zymogen granules of condensing vacuoles apparently formed by the Golgi complex. The inhibitors they used were: (a) cycloheximide, an inhibitor of protein synthesis, (b) antimycin A, an inhibitor of the electron transport chain and (c) sodium fluoride (NaF) to inhibit glycolysis.

condensation of enzyme within vacuoles

> **ITQ 3** (*Objective 6*) If the use of these inhibitors were intended to give information on the step CV → zymogen granule, how long after the administration of the pulse would it be reasonable to wait before applying the inhibitors—assuming their action was rapid?

> **ITQ 4** (*Objective 6*) What would the results obtained by the use of these inhibitors tell you of the condensation process?

By comparing the accumulation of label in the zymogen granules of slices (from the same animal) which had cycloheximide in the chase medium with the accumulation of label in controls in normal chase medium, Jamieson and Palade showed that further protein synthesis was not necessary for the conversion; it carried on regardless of the cycloheximide.

protein synthesis and condensation

The effect of respiratory inhibition was a little less clear-cut. After 20 minutes in the presence of antimycin A, and NaF there was no difference between the concentrations (in vacuoles and zymogen granules) in the experimental and control slices; both proceeded normally.* However, comparison at +60 minutes showed that there was about 40 per cent repression of conversion from vacuole to zymogen granule.

respiration and condensation

Thus, it is fair to say that conversion is to a large extent (at first completely) independent of respiration. Further experiments, with ouabain (see Unit 4,

* This effect was not due to the wrong selection of inhibitor; antimycin A and NaF in the same concentration almost completely (95 per cent) blocked the earlier step, that is, from RER → Golgi vesicles.

Section 4.3.2), confirmed that an active (sodium) pump of the kind commonly found in plasma membranes was not involved. Furthermore, none of the steps taken to poison an osmotic pump resulted in the swelling or dilution of existing zymogen granules. Clearly, the active extrusion of water is not a factor.

How then is the condensation achieved? Once again the answer is far from obvious. It may be that there is an 'internal' (to the vacuole, that is) process of aggregation of the protein molecules which does not require an outside energy source. It may be relevant to note that there is evidence that the enzyme is more readily released from condensing vacuoles by mild osmotic shock than it is from zymogen granules. Thus, when Jamieson and Palade (1971) shocked (with a 0.3 M sucrose solution) a fraction of the homogenate containing condensing vacuoles plus zymogen granules at +20 minutes, when the label was predominately in the CVs, 39 per cent of the protein was released, but at +80 minutes only 14 per cent was released. (Other treatments, e.g. dissolving the membrane in deoxycholate, give no grounds for supposing the membrane of the CVs to be significantly more fragile than that of the zymogen granules.)

This suggests that in the zymogen granules the enzyme is in larger aggregates, possibly making the contents less osmotically active than the contents of the CVs, which will take up fluid and burst in a hypotonic medium.

(e) *How are zymogen granules released from the cell?*

Since 1966 a number of electron micrographs have been made of various secretory cells which appear to show that zymogen (or other) granules move to the apical surface of the cell where their membrane fuses with the apical plasma membrane of the cell, thus releasing the contents outside the cell membranes. If we assume that the evidence for this process in the pancreas is conclusive, then this clearly forms the final step in the whole 'cisternal package' pathway for the export of enzymes. However, we pointed out earlier that the possibility of an alternative pathway for some enzymes in some species cannot be ruled out and this pathway could be through the cytoplasmic matrix and directly across the plasma membrane via specific carrier sites.

But whether or not we accept the 'cisternal package' route as the only one, it exists; and the release of the granules poses some interesting questions. One of these is how the movement of the zymogen granule to the apex of the cell (and the subsequent fusion) of the membranes is achieved.

We have already discussed the time course of the movement of labelled enzyme from the Golgi to secretion into the lumen of the acinar ducts in a fed animal. The interval between the peak of the pulse in the CVs and that in the zymogen granules (+57 to +117) is 60 minutes, and by +117 minutes secretion is already taking place, indeed 10 per cent of the visible autoradiographic grains have already been secreted (Table 3). Thus the time interval is comparable with the other stages of the process.

To clarify the release process Jamieson and Palade (1971) again treated their preparation with inhibitors (as they did when investigating the metabolism of the CVs), but this time they applied the inhibitors later.

The use of cycloheximide established that discharge did not depend on new protein synthesis; indeed they also showed that, after the original incorporation of the label into the cisternae of the RER, the whole cycle could occur in the presence of cycloheximide.

Unlike condensation however, it appeared that granule discharge *did* depend strictly on respiratory energy. Whether the energy is necessary for the movement of the granule, or for the process of fusion of the granular and plasma membranes, or both, is not known for certain.

There is evidence that if ATP is supplied to the cells from outside, it causes secretion of the zymogen granule contents and there is ATPase activity in the granule membrane. Furthermore, it can be shown that, during secretion there is an increased movement of Ca^{++} ions, from an intracellular 'pool' of some description, to the acinar lumen. This has led to speculation that the secreting granules may be drawn to the cell membrane by intracellular contractile microtubules, and that the involvement of Ca^{++} ions and ATP (and thus, of course, respiratory energy) may be similar to that in the contraction of smooth muscle

(see Unit 6, Section 6.7). Thus it could be argued that the respiratory energy is probably needed primarily to move the granules up to the cell surface. These ideas will be explored further in Unit 16.

Summary of Section 5.2.3

The pathway through the pancreatic acinar cell suggested by Palade and his co-workers is summarized in Figure 13. It is very simple to visualize the process, which is one of the reasons why his theory has a strong appeal. However, it clearly is not enough to recall merely the suggested pathway; it is important also to be aware of the evidence available to support the theory, and to know the sort of techniques by which the evidence was obtained. Without a grasp of the evidence and the methods, you will be in no position to evaluate this theory against others you may find in the literature, now or in the future.

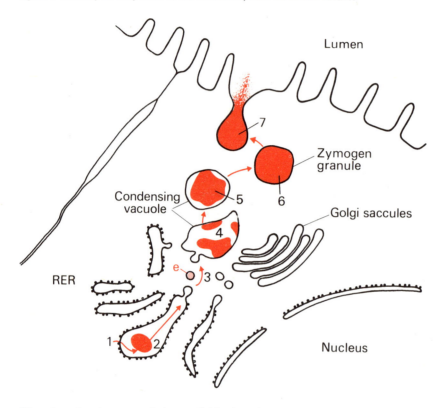

Figure 13 Summary of intracellular pathway of exportable enzyme.

We selected only some of the available data for presentation in Section 5.2, but SAQs 1–5 on p. 38 take the steps numbered 1–5 in Figure 13 and ask what kind of evidence was quoted to support each one. These self-assessment questions constitute the best summary of this Section; *you should attempt them now.*

5.3 Secretion of electrolytes and water by the pancreas

The normal secretion produced by the mammalian pancreas after a meal has been eaten includes Na^+, K^+, HCO_3^-, Cl^- and, of course, water in addition to the enzymes. The Na^+ and K^+ are present in relatively constant concentrations, these concentrations depending on the constitution of the fluid supplying the gland—normally, of course, the blood plasma. The concentrations of the principle anions (HCO_3^- and Cl^-) vary, however, but the sum of the two remains constant and approximately the same as that of the cations. Put another way, when the HCO_3^- falls the Cl^- rises by a comparable amount, the ratio of anions to cations remaining more or less constant.

QUESTION   Is it of importance that there should be a high level of HCO_3^- in the pancreatic juice when it is being secreted rapidly?

ANSWER   Yes, because high flow rates of juice coincide with the digestion of a meal[4], and during digestion the very acid stomach contents enter the small intestine. The stomach acid must be buffered, as the pancreatic and intestinal enzymes work at a much higher pH. The pancreatic juice is a major buffer in this system.

22

An examination of the exocrine pancreas under the light microscope (Figs. 1 and 2 on p. 9) might lead to the suggestion that probably both the enzymes and the electrolytes are produced by the acinar cells and secreted into the acinar lumen; the secretion would then pass along the intralobular ducts into the interlobular duct and then into the main collecting duct. However, some of the facts mentioned in S22–4 and earlier in this Unit suggest that different cells may be involved in the secretion of protein on the one hand and electrolytes and water on the other.

ITQ 5 (*Objective* 7) What information suggests that different cells may be involved?

It has already been suggested in this Unit, on the basis of some of this evidence, that it may be the centro-acinar cells and the lining of the duct system that are primarily responsible for the fluid and electrolyte content of the pancreatic juice.

5.3.1 How is the primary secretion produced?

It might be possible for the flow of water and electrolyte to be derived from the blood plasma by a process of ultra-filtration—in much the same way as the primary secretion is produced by the glomeruli of the kidney.[8] The 'sieve' would have to be finer than in the kidney to hold back amino acids, sugars, etc., or the ducts would have to be capable of reabsorbing them, in a manner analogous to the tubules of the kidney.

This cannot be entertained as a serious hypothesis for a number of reasons: for example, there are none of the other constituents you would expect to find in a plasma filtrate, such as urea, phosphate, etc., and there are no structures in the acini (see Fig. 1 on p. 9) that remotely suggest a pressure filtration system. Nor is the production of juice related to blood pressure as it is in other ultra-filtration systems. So we are looking at a system in which water and electrolytes are secreted either by a process of selective diffusion down an electrochemical gradient, or by the active transport of one, or several, of the constituents.

It has been shown that, in fact, secretion of the juice is dependent on respiratory energy. If the pancreas is isolated, and perfused, either through its own vascular system or bathed *in vitro*, secretion will cease if the preparation becomes anoxic, or if it is treated with cyanide or 2,4-dinitrophenol. Removal of glucose from the perfusate seriously diminishes secretion, though it can be maintained with pyruvate or lactate. Thus, it appears that at some stage the production of the primary secretion is 'driven' by oxidative phosphorylation.

The current hypothesis on the origin and constitution of the secretion, as it is first produced by the cells (the primary secretion), derived from the slightly surprising relationship mentioned earlier between the HCO_3^- and Cl^- ions in the final juice. It was known that the *sum* of the anion concentrations remained roughly constant at both high and low flow (i.e. secretion) rates, but that the concentration of HCO_3^- was inversely proportional to the flow rate whereas the Cl^- concentration was directly proportional to it* (see Fig. 14).

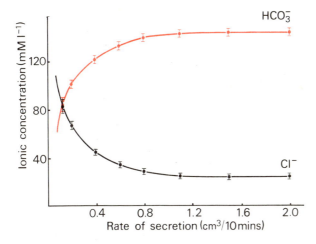

Figure 14 The concentrations of HCO_3^- and Cl^- ions in the pancreatic juice at different rates of secretion.

* These statements apply mainly to a range of laboratory animals, on which most of the experimental work has been done. The situation in man is rather less clear because the results are based on duodenal sampling, as it is not possible to retain a cannula in the pancreatic duct of a human subject without causing damage.

23

It was a reasonable argument that the juice reaching the intestine would be more nearly like the primary secretion at high secretion rates than at low ones on the grounds that there would be less time to modify the primary secretion, and by analogy with other secretory systems such as the kidney. (It is possible, of course, that the primary secretion itself is different at different flow rates, and that this relationship between the anions is not due to subsequent modification; however, it is harder to visualize a secretory mechanism in which this would be the case.)

Thus, it was postulated that the central secretory event is the active secretion of bicarbonate, HCO_3^-, probably by the centro-acinar cells. To ascertain whether the cations (notably Na^+) would need to be actively extruded to account for the concentration found in the juice, certain measurements would have to be made.

active secretion of HCO_3^- by centro-acinar cells

> QUESTION (*Objective 7*) What information is required to make this judgement?

> ANSWER   We should need to know the concentrations of electrolytes inside and out (see Unit 3, Section 3.6.1), and the potential difference generated between the inside of the cell and the fluids either side of it as a result of the distribution of ions.

It appears that the potential difference between the cells of the acini* and the plasma (or perfusate) and the relative concentrations of electrolytes are such that Na^+ could enter the base (contraluminal side) of the cell from the plasma down the electrochemical gradient. However, if the same gradient exists between the cells and the acinar lumen, then clearly Na^+ would have to be pumped from the cell across its luminal surface, though it is possible that the active extrusion of HCO_3^- during secretion might produce an electrical gradient sufficiently 'steep' to allow the passive exit of Na^+. Such a gradient was not found by Case *et al.* (1969), and they concluded that it might well be that the Na^+ was carried across the luminal membrane by the same carrier system as the HCO_3^-, against its electrochemical gradient (see Fig. 15).

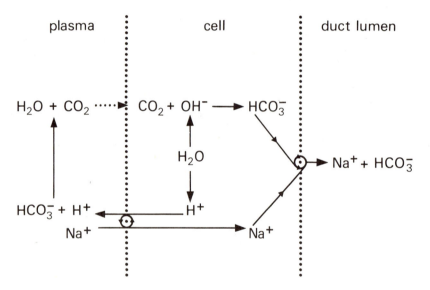

Figure 15 Scheme of events in the secretion of HCO_3^-.

The source of the bicarbonate secreted in the juice is also a matter of some interest. Different workers have variously considered that it is derived from the metabolic CO_2 of the gland cells, from the CO_2 in the blood or the HCO_3^- in the plasma.

source of HCO_3^-

There is still some confusion, but it now seems fairly certain that some 5 per cent of this CO_2 arises from the metabolic activity of the cell itself, but the remaining 95 per cent is derived from the plasma, mainly the plasma HCO_3^-. Thus the removal of HCO_3^- from the perfusing fluid severely reduces bicarbonate secretion by the isolated gland. (In naturally secreted juice the HCO_3^- concentration is about four times that in the plasma, and it is interesting to note that this is also

---

* This is, of course, a dubious piece of evidence in this context, as the *acinar* cells are probably not the site of electrolyte secretion, and the measurement has never been made on the centro-acinar or intralobular duct cells.

true in the perfused gland—the juice produced has about four times the HCO_3^- concentration of the perfusing fluid.) It appears that the gland does not take up the plasma bicarbonate as such, but as CO_2, which diffuses in. The gland cell secretes H^+ into the blood (in exchange for Na^+) which produces the following reaction:

$$H^+ + HCO_3^- \rightarrow H_2CO_3 \rightarrow H_2O + CO_2$$

The enzyme carbonic anhydrase, within the cell, takes a part by hydrating CO_2 to form bicarbonate. This part of the process can be summarized as shown in Figure 15.

On this basis the main activity of the cell at its contra-luminal surface, in connection with bicarbonate secretion, would be the ionizing of water, $H_2O \rightarrow OH^- + H^+$, and the subsequent secretion of H^+.

ionization of water

5.3.2 How is the primary secretion modified?

If the theory outlined above is correct, at least so far as it applies to laboratory animals, then a copious flow of fluid is produced (during secretion) with the constituent electrolytes being mainly Na^+ (approximately 160 mM l^{-1} and HCO_3^- (140 mM l^{-1}, but also with much lower concentrations of K^+ and Cl^- (approx. 5 and 25 mM l^{-1} respectively).

It may be that after a meal, with the maximum pancreatic flow being stimulated by secretin and pancreozymin, little or no modification of the primary secretion takes place. However, at lower flow rates there are substantial changes.

Case *et al.* (1969) performed experiments on cat pancreas, to see if the ducts modified the secretion passing through them. The cat is a particularly suitable animal as it has a long, straight main pancreatic duct, through which test solutions can be run. They inserted one cannula at the glandular end of the duct and collected from another one at the duodenal end. They then irrigated the duct at different flow rates either with natural pancreatic juice (collected from another cat at maximum flow rates) or with an artificial solution containing appropriate concentrations of the electrolytes. The results they obtained (Fig. 16) bear a striking resemblance to the results shown in Figure 14 (on p. 23), which show the changes in the natural juice arising from changes in the flow rate *in vivo*.

perfused pancreatic duct

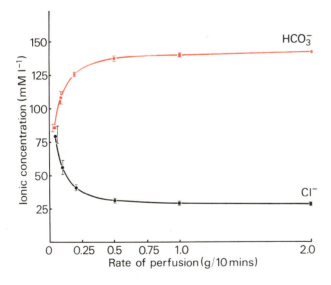

Figure 16 The concentrations of HCO_3^- and Cl^- ions emerging from the perfused pancreatic duct of a cat, at different rates of perfusion.

These results lend very strong support to the view that the primary secretion is basically a Na^+, HCO_3^- solution, but that the ducts modify it by absorbing HCO_3^- and adding Cl^-. The slower the rate of secretion, the more this process occurs.

However, this view is not unchallenged. Shultz (1969) performed micropuncture studies on the rabbit. Figure 17 illustrates the techniques.

Shultz introduced a droplet of coloured oil into an interlobular duct with a micropipette, and then 'split' the droplet with a drop of the solution to be tested, so that it was trapped in the duct between coloured droplets of oil. She

micropuncture techniques

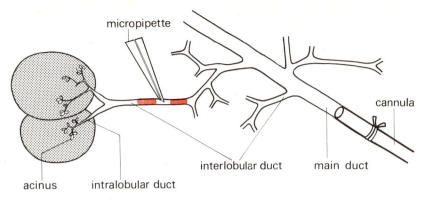

Figure 17 The 'split oil droplet' micropuncture technique used by Shultz.

could then remove and analyse the drop, and by repeating the process further down the duct, she could compare the activity of different regions. She also took samples of the natural juice from different points, and compared them with the samples obtained from the cannulated main duct.

Using both these methods, Shultz concluded that the Cl⁻ concentration *decreased* steadily as the fluid passed from the interlobular ducts to the main duct, at least in the stimulated gland. She also compared the concentration change of Cl⁻ with that of the large carbohydrate molecule, inulin, which she introduced with the drop (inulin is too large to cross the duct epithelium). From this comparison she concluded that water was added to the lumen by the ductal epithelium. Her conclusions were therefore that water and HCO_3^- were added continuously to the secretion throughout the length of the duct system in the stimulated gland of the rabbit. The implication of this is that the 'primary acinar secretion', if it exists at all, may be merely a small, chloride-rich flow. Alternatively, it may mean that both secretion and modification occur to a varying extent in all the small ducts—and that centro-acinar cells are really only a part of the intralobular duct system.

You should now attempt SAQ 6 on p. 38.

5.4 Control of pancreatic secretion

In view of what has been said in this Unit about the separation of enzyme secretion from the secretion of the bulk of the fluid and electrolytes, and in view of the fact that at least two different gastro-intestinal hormones are involved in stimulating these secretions, it is clearly necessary to consider separately the control of the secretion of these two components of the juice. However, some observations are common to both. The secretion of pancreatic juice is discontinuous and determined by the animal's feeding habits; so neither the aqueous nor the protein components are secreted continuously. There is no evidence to suggest that any cells manufacture and store any part of the aqueous component when the gland is unstimulated, but enzyme precursors are made and stored in the acinar cells in the absence of stimulation.

It also appears that the view, expressed in Section 5.2, that secretin should be considered the stimulus for water and bicarbonate secretion, and pancreozymin the stimulus for enzyme release, is an oversimplified one; there are inter-hormonal effects. Secretin and pancreozymin are released from the wall of the small intestine, principally as a consequence of the arrival of food and gastric juice in the duodenum. Secretin release is dependent on the pH of the acid, pancreozymin release on the presence in the duodenum of various essential amino acids and some fatty acids. However, the effect of secretin in promoting electrolyte and water secretion by the pancreas is much increased if the pancreatic branch of the vagus nerve is first stimulated, or the cells are treated with acetylcholine. In turn, pancreozymin produces a much greater enzyme release if the gland has been pre-treated with secretin, and the effect of secretin on the release of electrolyte is enhanced by pre-treatment with pancreozymin. Thus, although there are separate discernible roles for the vagus nerve, secretin and pancreozymin in producing pancreatic secretion, it is clear that there must be functional overlap, to produce these effects of 'reinforcement'.

5.4.1 Enzyme secretion

How pancreozymin acts to cause the release of zymogen granules is not yet known—it may well be soon, but the vital step is still missing. It is known that protein synthesis is not necessary but that respiratory energy is. Jamieson and Palade (1971) showed that pancreozymin did not speed up the rate of movement of exportable enzyme through the cell in the earlier steps of the process so it must act on the kinetics of the last step only. Pancreozymin is a polypeptide; you are aware from earlier Courses[6] that polypeptide hormones such as vasopressin can act via the 'second messenger', cyclic AMP, to bring about the production of intracellular enzymes that in turn affect the kinetics of various reactions. These facts together lead to all kinds of exciting speculation, but, as we said above, the vital link is missing, and so far there is no convincing evidence that cyclic AMP is involved in enzyme release.

stimulation of the acinar cells by pancreozymin

Finally, a real teaser. You will recall Rothman's work on enzyme 'profiles' (see p, 14), that is, on the ratio of one enzyme to another in different parts of the intracellular pathway. He measured (1969) the following ratios in the rabbit (see Table 4).

Table 4   Comparison of the ratios of Tg and ChTg during secretion by the rabbit pancreas

Fraction	Ratio of Tg : ChTg
Microsomes	2.9 : 1
Zymogen granules	2.6 : 1
Secretion	6.1 : 1

When he administered pancreozymin before making the measurements, the difference between the ratios in the zymogen granules and the secretion was even greater (Table 5).

Table 5   Comparison of the ratios of Tg and ChTg after stimulation by pancreozymin

Fraction	Ratio of Tg : ChTg
Microsomes	2.8 : 1
Zymogen granules	2.6 : 1
Secretion	15.8 : 1

ITQ 6   What are the implications of these figures?

5.4.2 Water and electrolyte secretion

Whether the production of water and electrolyte depends on a primary secretion rich in bicarbonate from the acini, which is modified by the addition of chloride in the ducts, or whether the secretion is primarily ductal, the control mechanism is likely to be broadly similar. In either case it is likely that the active pumping of H^+ into the plasma, and HCO_3^- (and possibly Na^+) into the lumen are central events, and that water and the other electrolytes follow. Thus, we are considering what link there could be between the arrival of secretin in the plasma and the activation of an energy-consuming ion pump, presumably situated on the luminal membrane of the cell.

Case *et al.* (1969) (also Scratcherd and Case, 1973) have reported a sudden transient escape of K^+ ions into the fluid perfusing the acini when the gland is treated with secretin. This has lead them to speculate that secretin may act on the contraluminal surface of the centro-acinar cells to increase its permeability. Then K^+ might escape down its electrochemical gradient into the plasma and Na^+ would enter. This ionic shift might set in motion some stage of the process summarized in Figure 15 (p. 24). However, the permeability change may not

secretin and cell permeability

be a primary or essential step in triggering secretion; it has now been established beyond reasonable doubt that cyclic AMP *does* act as a second messenger between secretin and the secretory mechanism. What the connection between cyclic AMP and the ion pumps is, remains to be seen.

5.5 Milk secretion

There are many important physiological differences of which you are already aware between the exocrine pancreas and the mammary gland. Nevertheless, in terms of the secretory mechanisms, it is reasonable to expect that there may also be similarities: the pancreas, at maximum secretion, produces a juice made up predominantly of water, some electrolytes and protein enzymes—and milk is predominantly water (80–90 per cent in domestic animals) and includes electrolytes and protein. Therefore, it is interesting to look at the secretion of these components to see if they are produced by a common mechanism or by a different one.

There are, of course, important constituents *not* found in pancreatic juice, the two most notable being sugar, in the form of lactose (as much as 6–7 per cent in women and mares) and fat (virtually nil in the rhinoceros, about 4 per cent in the cow and up to nearly 50 per cent in seals and other aquatic mammals). The secretion of these substances obviously has to be looked at in terms of possible 'new' mechanisms.

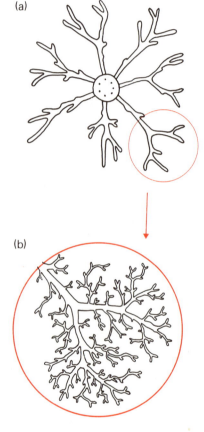

5.5.1 Origin and structure of the mammary glands

In a previous Course[3] we discussed a little of the hormonal background to the development and maintenance of the mammary glands, but said nothing about their origin and structure—we conjured them out of thin air! In reality they appear in a rather different manner; there is good evidence that they develop from sweat glands in the skin. There are areas of the mammalian ventral surface known as the 'milk lines'—areas of skin, including skin in the axillary regions (i.e. armpits and groin) and two strips connecting these areas, running over the chest and abdomen. The sweat glands in these areas appear to function rather differently from those on the remainder of the skin; for one thing their secretion contains a good deal of protein.

The secretions of these glands are the life blood, so to speak, of the deodorant industry of the Western world, as their constituents are liable to bacterial decomposition, as well as having an intrinsic smell.

In many mammals, some of these glands become modified to produce scents which have considerable sexual importance, as pheromones.[7] More important perhaps, it appears that sweat glands along the milk line may also become specialized as mammary glands, at least in placental mammals.

The glands are derived during development from infoldings of the epidermis of the skin which form branched blind-ending ducts. These are present in both sexes.

> QUESTION Can you recall a difference between the process of mammary gland formation and that of the formation of the pancreas?

> ANSWER In mammary glands the ducts are formed from *epidermal* tissue; in the pancreas they are outgrowths of the *endodermal* lining of the gut. This difference might result in differences both in structure and function of the duct cells.

The deposition of fat to form the general shape of the mammary gland takes place at puberty, under the influence of oestrogen. When the animal becomes pregnant the ducts grow and branch into the fat under the influence of oestrogen and progesterone[3] and eventually the terminations of the ducts form the secretory alveoli (see Fig. 18).

Morphologically the glands may be of various types, but in any type a so-called 'gland' is, in fact, hundreds of separate glands, each with a lactiferous duct,

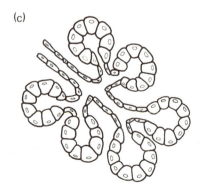

Figure 18 The arrangement of mammary ducts: (a) main ducts branching from the teat; (b) fine ducts branching from the main ducts and (c) fine ducts terminating in 'branches' of secretory alveoli.

28

and a lactiferous sinus of its own. These sinuses may open directly into the teat, when they are small (as in women, bitches and sows), or they may be expanded to form a relatively large 'milk cistern' (as in cows and goats). Over what regions of the milk line mammary glands form depends on the species; it may be just in the inguinal region (cow) or just in the thoracic region (primates), or all the way along (sow).

5.5.2 Secretion of milk proteins

The principal proteins found in milk are casein, α-lactalbumin, β-lactoglobulin and to a lesser extent, various immunoglobulins (γ-globulins). The significance of the immunoglobulins to the neonatal animal will be discussed in Unit 11; their importance in terms both of the proportion of the total secreted protein that they comprise, and their physiological function, is most marked when the gland is producing colostrum[3] rather than when it is in full lactation. Evidence has now accumulated which shows that most of the immunoglobulins are derived directly from the plasma, rather than synthesized from smaller precursors by the secretory cells of the gland, as is the case with the casein, lactalbumen and lactoglobulin. It is logical therefore to deal with the secretion of these two groups of protein separately.

origin of milk immunoglobulin

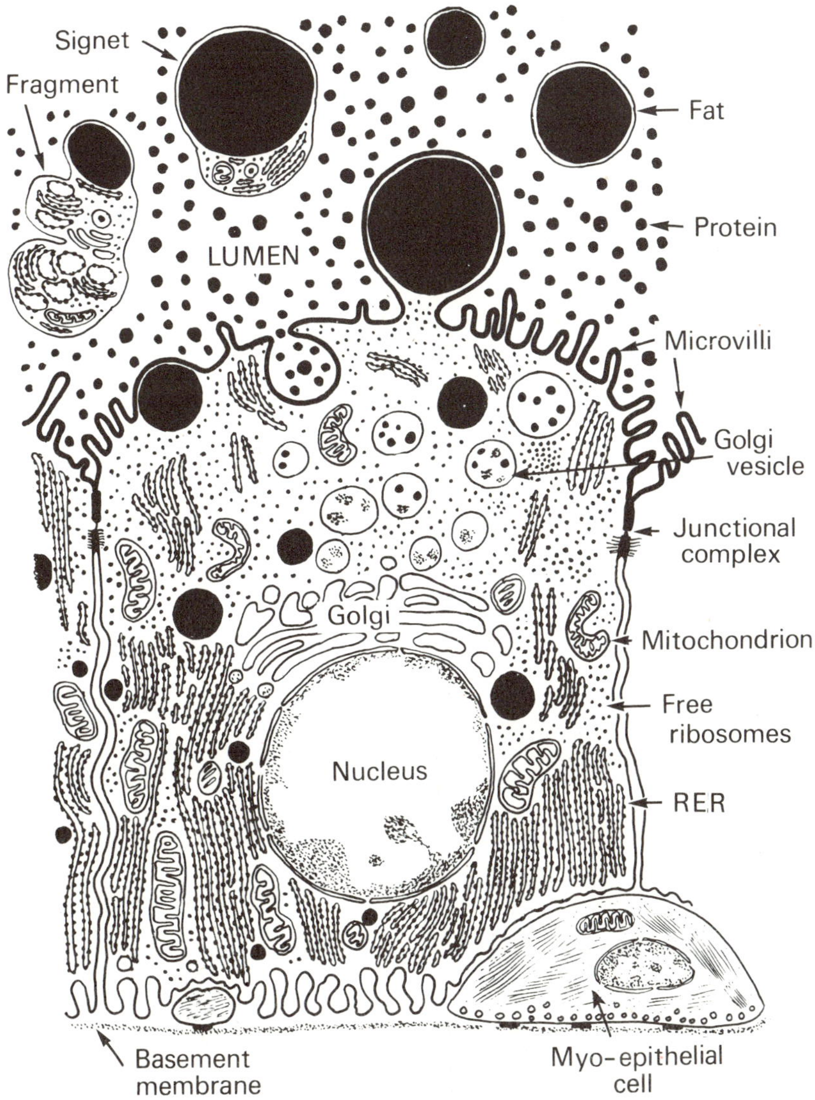

Figure 19 An alveolar cell secreting milk (reconstructed from electron micrographs).

mammary alveolar cell

Figure 19 is a diagram drawn from electron micrographs, showing the structure of the secretory alveolar cell. Much of the general internal arrangement is reminiscent of the pancreatic acinar cell: (a) the large round nucleus is towards the base (contra-luminal side) of the cell; (b) the basal part of the cell is filled with tightly-packed RER showing a 'preferred orientation'; (c) there is a large Golgi complex containing vesicles above the RER and nucleus, and larger vesicles can be seen further up towards the apex of the cell.

Granules can be seen with the electron microscope in the Golgi vesicles, larger apical vesicles and lumen of the alveoli. Pre-treatment of the material with trypsin removes these granules. EM autoradiographic techniques, essentially similar to those used by Palade and his co-workers, have shown the accumulation of labelled amino acids in the RER, Golgi vesicles, larger apical vesicles and lumen of the alveoli.

This evidence gives reasonable grounds for believing that protein secretion, or much of it at least, takes place along a cellular pathway basically similar to that in the exocrine pancreas.

However, the story is less fully worked out than it is in the pancreas: the step which takes place between formation in the RER and incorporation into Golgi vesicles is subject to speculation, and less thorough biochemical analysis of the early stages has been undertaken. Kimura (1969) and Kurosumi *et al.* (1968) are of the opinion that the protein is not fully formed until a 'post-Golgi' vesicular stage, and that the earlier granules are smaller polypeptides. If this is correct, it might constitute a difference between the pancreatic and mammary mechanisms.

polypeptide granules in RER cisternae

The secretion of immunoglobulin by the gland, particularly in the colostrum, presents a rather different picture. Studies using labelled amino acid precursors or immunoglobulin in the blood plasma (the techniques of various types of mammary study are examined in the TV programme associated with this Unit) show that these particular proteins are not manufactured *de novo* in the secreting cell from amino acids, but whole from the blood plasma.

Some workers have assumed for many years that the immunoglobulins get from the plasma into the alveoli by what amounts to a passive leak, which occurs particularly during the pre-milking phase when colostrum is formed and there is a high pressure in the alveoli. As techniques were developed for the precise separation and identification of proteins it became apparent that a passive leak could not account for the situation.

Although the proportion of the immunoglobulins removed from the plasma into the alveoli is sufficiently large to produce a measurable drop in the total plasma protein level, the 'profile' of the immunoglobulins in the alveoli and the blood (or extra-cellular fluid) is not the same, that is, there has been selective transfer. What is more, this selective mechanism can be shown to distinguish even between two immunoglobulins of closely similar weights.

QUESTION   What sort of mechanism does this suggest to you?

ANSWER   A special carrier site—though on the information given, it may not involve the expenditure of energy.

The glands of different species select different immunoglobulins for transport into the colostrum or milk. For example, in the cow, sow and ewe, it is mainly immunoglobulins G and M (see Unit 11), whereas in women and rabbits it is immunoglobulin A.

The immunoglobulins are synthesized by cells called 'plasma cells' (see Units 8 and 11). These originate in the bone marrow and are released into the blood circulation as white blood cells (lymphocytes in fact) which later become plasma cells. These may migrate out of the circulation into the extra-cellular fluid, or remain in the general circulation. If a foreign protein (antigen) comes into contact with them, they will manufacture and release a specific immunoglobulin (antibody) which will split or agglutinate the antigen.

Thus the immunoglobulins transported into the alveoli might either have been picked up from the general circulation as the plasma passed through the gland or have been taken up from plasma cells which settled in the extra-cellular fluid in the gland. There is evidence (Lee and Lascelles, 1969; McDowell and Lascelles, 1969) that the latter is the case with immunoglobulin A in sheep. If antigens are introduced into the mammary gland on one side of a sheep, it can be shown that only that gland, and not the one on the other side, secretes the appropriate antibody. Furthermore, only the side on which the antigen was introduced shows a visible increase in the number of plasma cells in the tissue spaces.

30

Finally, it has been shown in the cow that whereas both immunoglobulins and albumen can be detected in the extra-cellular space outside the alveolar cells throughout all stages of lactation, during colostrum secretion large quantities of the immunoglobulin and none of the albumin can be shown to be inside the alveolar cells and lumen.

There is no doubt therefore that the immunoglobin fraction of milk and colostrum protein is derived not from the alveolar cells but from plasma cells, and that its transport into the alveolar lumen is selective, taking place through the alveolar cells. There is some evidence that the immunoglobulin is taken up by a process of pinocytosis at the base of the alveolar cell.

selective transport of immunoglobulin

There is no evidence that, once in the cell, the immunoglobulins enter the Golgi vesicles; it seems probable therefore that they pass through the cytoplasmic phase rather than the ER cisternae, and that their secretion into the lumen depends again on specific carrier sites.

You should now attempt SAQ 7 on p. 38.

5.5.3   Secretion of milk fat

There are certain risks inherent in attempting to generalize across the species about the processes whereby milk fat is synthesized and secreted, and one of these risks results from the wide variation in fat content (from almost nil in the rhinoceros to 50 per cent in the seal). There is quite a possibility that very large scale secretion will involve different pathways and different precursors from those involved in more modest production. Also, by far the greatest amount of work has been performed on the rat and the cow, and more recently, the goat. Really, very little is known about fat secretion in any of the species with very high or very low levels of milk fat.

What has been established, however, is very interesting. Mammary tissue *in vitro* can manufacture fatty acids, through a reaction which requires the consumption of glucose. In the rat, carbon from the glucose molecules can be incorporated in the carbon skeleton of the fatty acids, but in ruminants (cow, sheep, goat, etc.) this is not so; the fatty acids must be built up using substances (acetate, hydroxybutyrate) released into the plasma by the fermentation processes occurring in the rumen. (This results in the production of fat with a high proportion of short-chain fatty acids (with 4–14 carbon atoms), giving butter its distinctive taste; you might *not* be able to tell mouse butter from Stork!)

sources of fat in ruminant milk

Linzell (1968) has been able to quantify the precursors of the fats found in goat's milk, and has shown that the majority of the short-chain fatty acids found in the milk fat are derived from fat droplets (chylomicrons), and a little of the glycerol, found in the plasma. Most of the longer chain ones are derived from free fatty acids and triglyceride in the plasma. In the goat the proportion of milk fat built up in the gland cells from non-lipid sources is 40 per cent and mainly confined to the short-chain fatty acids.

The process by which the alveolar cell is able to make use of the chylomicrons in the blood is truly remarkable. Chylomicrons are large by cellular standards (often 1 μm in diameter); they appear in the blood after a fatty meal, and are released from the lymphatic vessels draining the intestinal villi, sometimes in such large numbers that the blood takes on a milky appearance to the naked eye. It is hard to imagine how such large droplets could leave the capillaries and enter the alveolar cell. Furthermore, EM pictures have failed to show large droplets entering the basal areas of the alveolar cell.

In 1967 Schoefl and French showed that chylomicrons appeared to be 'sticking' to the thin endothelial lining of the mammary capillaries, and that long processes were being put out from the cytoplasm of the endothelial cells which then surrounded the chylomicron (Fig. 20). It also appeared that the droplet as such did not cross the cell, but was split by some lipolytic enzyme and transported across the endothelial cell and the base of the alveolar cell as fatty acid.

uptake of chylomicrons from the plasma

This process of chylomicron uptake can be extensive; under some conditions the goat udder can remove up to 60 per cent of all the chylomicrons from the blood in a single passage through the gland.

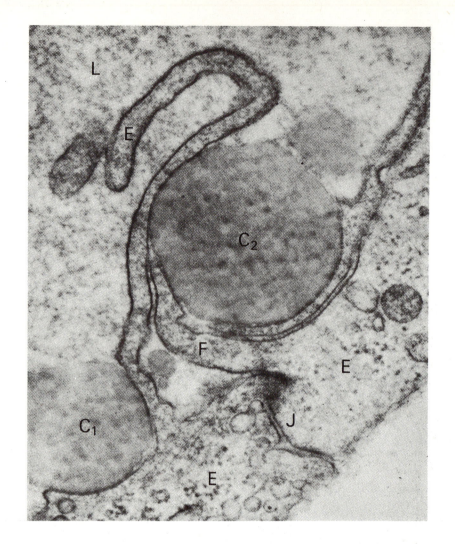

Figure 20 Electron micrograph showing two chylomicrons stuck to endothelial cells of a capillary in the mammary gland of a rat and being surrounded by processes from the cells. C_1 and C_2—chylomicrons, E—endothelial cells, F—processes of the endothelial cells, J—junction between two endothelial cells, L—lumen of capillary. ($\times 107\,000$)

The next main point to consider is where the fatty acids are joined together (esterified) to form the large fat molecules. The evidence from electron micrographs is that the Golgi apparatus is not involved, at least not directly. Some workers (Kurosumi *et al.*, 1968) basing their conclusions on the interpretation of EM pictures, believe that smooth endoplasmic reticulum (SER) forms vesicles in which the fat forms, but this is not a majority view. Using pulse labelling and autoradiographic techniques (very similar to Palade's on protein secretion), it has been shown in mice that esterification occurs very quickly in the cisternae of the RER, (within 1 minute) and then the label moves into the fat droplets in the

intracellular pathway of fat

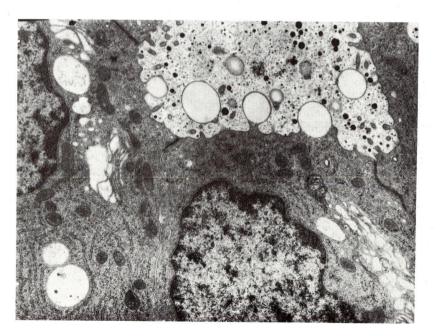

Figure 21(a) Secretion of fat droplets from a mammary alveolar cell. ($\times 10\,000$)

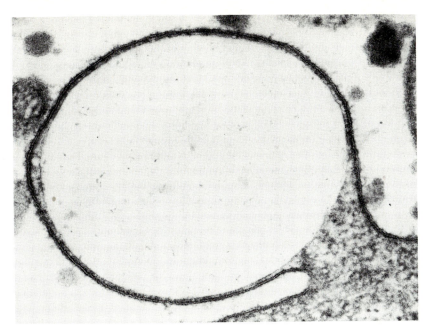

Figure 21(b) Note the cell membrane stretched around the droplets as they are extruded. ($\times 120\,000$)

cytoplasm and thence into the alveolar lumen. However, in goats, fat production does not seem to be confined to the cisternae—so perhaps there is a difference of the sort mentioned above, between animals with a higher and lower milk fat.

The mechanism by which the vesicles are extruded is perhaps the most fascinating part of the whole story. After years of controversy, there is now considerable evidence that as the fat droplet approaches the apical cell membrane, surface forces acting between the droplet (which is not surrounded by a membrane) and the membrane cause the latter to be drawn very tightly round the droplet and so-to-speak 'pinch itself off' from the rest of the membrane. Theoretical calculations indicate that the forces involved would be ample to do this. Thus the fat droplet would be secreted wrapped in apical cell membrane (see Fig. 21).

secretion of fat droplets

It has been shown chemically that milk contains the substances which make up the alveolar cell membranes, in quantities sufficient for this explanation to be possible. Furthermore, Wooding *et al.* (1970) showed that about 5 per cent of the fat droplets in goat's milk were surrounded by not only membrane but a 'signet ring' of cytoplasm from the alveolar cells as well (see Fig. 22).

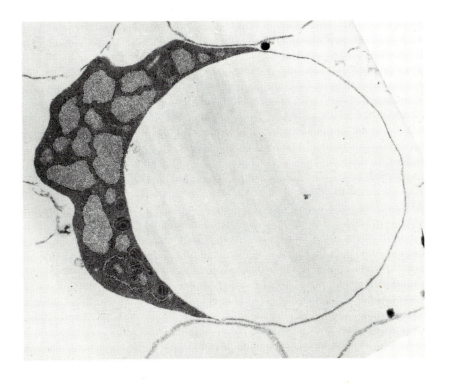

Figure 22 Fat droplet floating in goats' milk. The droplet is surrounded in apical cell membrane; in addition a portion of the alveolar cell has become 'pinched off' with the droplet, and shows mitochondria in transverse section and swollen ER. In the goat, this appears to happen with about 5 per cent of the secreted droplets. ($\times 18\,000$)

You should now attempt SAQ 8 on p. 38.

5.5.4 Secretion of lactose

The major carbohydrate constituent of milk is lactose. This is manufactured in the alveolar cells from glucose. The final step in this reaction, UDP-galactose + glucose → lactose + UDP, is catalysed by the enzyme, lactose synthetase. It appears (Brodbeck *et al.*, 1966) that this enzyme is present in two parts, the A and B proteins, and that the B protein is in fact α-lactalbumin, which you may recall has long been known as a protein constituent of milk, comprising in fact, some 12 per cent of the total milk protein.

lactose synthetase

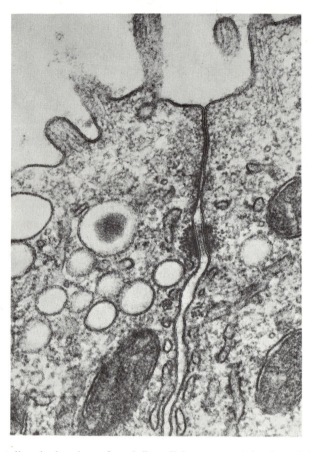

Figure 23 Electron micrograph showing tight junction between two alveolar cells. ($\times 77\,000$)

The α-lactalbumin has been found (in cell homogenate) in the soluble cytoplasmic fraction and in microsomal ones. It has also been shown to occur in the cisternae of the RER.

The A protein, on the other hand, is found bound to microsomal fractions believed to be part of the Golgi apparatus, and Golgi vesicles have been isolated containing both A protein and α-lactalbumin.

> ITQ 7 Do these data suggest to you a possible site and pathway for lactose synthesis and secretion?

The hypothesis that the lactose is manufactured in the Golgi vesicles receives additional support from Linzell and Peaker (1971) who have shown that the apical surface of the alveolar cells is surprisingly highly impermeable to lactose; lactose introduced via the teat does not enter the cells or the blood. Indeed, labelled lactose does not enter the alveolar cells from the plasma, but remains in the extra-cellular fluid. Thus it was very difficult to imagine how the lactose escaped from the cell into the alveolar lumen, until Brew's suggestion (see the answer to ITQ 7) that it was secreted in the vesicles. In a sense, the interior of the vesicle is already 'outside' the cell, so on this basis the lactose is actually made outside the cell membrane.

5.5.5 Secretion of water and electrolytes

Milk is isosmotic* with plasma, but as lactose is the major osmole, it follows

---

* ISOSMOTIC: Exerting equal osmotic pressure.

that the electrolytes (Na^-, Cl^-, Ca^{++}, Mg^{++}, HPO_4^- and citrate) are present in lower concentrations than in the plasma.

It has been held that the water and electrolyte component of milk has a mixed origin; that a secretion is produced by the alveolar cells, along with the protein, etc., and this is 'diluted' by a leak of plasma fluid between the cells. This was held to account for the plasma proteins (mainly immunoglobulin) which can be found in milk, and particularly in colostrum.

ITQ 8 What do you know about the secreted immunoglobulin which makes this explanation unlikely?

The morphology of the gland, as seen under the electron microscope (see Fig. 23) also makes a leak unlikely. The alveolar cells are joined by 'tight junctions' towards their apex; experience with other tissues suggests that molecules as large as protein seldom, if ever, escape past such junctions.

An early step, when examining water and electrolyte secretion in any tissue is to ascertain which movements of ions are apparently active, and which show the expected passive distribution, given the electrochemical gradients. You are, by now, thoroughly familiar with this exercise, and we suggest that you do what Linzell and Peaker (1971) did after they had made the necessary measurements, that is, calculate the equilibrium potentials by applying the Nernst equation (see Unit 3) to the Na^+, K^+ and possibly Cl^-, in the extra cellular fluid (ECF), alveolar cell and milk (Table 6).

Table 6 Distribution of electrolytes between the alveolar cell, ECF and milk

	ECF	Alveolar cell	Milk
$[Na^+]$/mM	150	43	8
$[K^+]$/mM	4.5	143	24
$[Cl^-]$/mM	116	62	12
Potential difference/mV	Taken as reference = 0	−41 (PD$_{obs}$)	+3 (PD$_{obs}$)

ITQ 9 (*Objective 10b*) Taking the distribution of ions across the *basal* surface of the cell first, what can you conclude, by calculating the equilibrium potentials of Na^+, K^+ and Cl^- between ECF and the alveolar cell?

ITQ 10 (*Objective 10b*) Now repeat the exercise for the apical membrane. What do you conclude from this?

Thus the situation appears to be that active regulation of the cellular electrolytes is occurring at the base (and perhaps in the intercellular spaces below the tight junctions).

As the apical surface is in equilibrium with the milk, so far as Na^+ and K^+ are concerned, the effect is that the basal part of the cell is potentially able to regulate these ions in the stored milk also.

Na$^+$ pump on the base of cells

QUESTION If whole cells, or even just the tight junctions, become damaged by bacterial attack (as in mastitis*), what would be an immediate effect on the constitution of the milk?

ANSWER Plasma electrolytes would enter the milk, raising the Na^+ and Cl^- level. Also, plasma proteins might leak in, and lactose out. This rise in milk Na^+ and Cl^- may become the basis for the simplest and most sensitive test for mastitis so far devised.

What then causes the movement of water from the plasma into the milk? As there is no Na^+ pump operating into the milk from the cell (and anyway, the electrolyte concentration of the milk is lower than that of the plasma), there is no

---

* MASTITIS: inflammation of the mammary gland; it leads to considerable economic losses, as milk from animals suffering from this disease cannot be sold.

obvious reason why the electrolytes should be considered a prime factor in the secretion of large amounts of water.

> QUESTION Can you think of any osmotically active milk constituent that might carry water with it into the lumen?

> ANSWER Lactose. We said at the beginning of this Section that it was the main osmole in milk.

Thus it might be that the water does not pass directly over the apical cell membrane at all, but is drawn by osmosis into the vesicles in which the lactose is synthesized, and released with the lactose and protein into the alveolar lumen. This could provide an explanation for the fact (Fig. 18) that the vesicles appear larger and more filled with fluid than their protein contents would justify.

link between lactose and water secretion

There is further support for this idea in the fact that in a perfused udder (see the TV programme for Unit 5) glucose is essential for the production of the aqueous part of the milk. Other sugars, such as mannose, which will act as energy sources can be used instead of glucose, but the gland then only produces an ooze, very rich in protein and fat. Clearly, without the glucose there will be no lactose, and apparently without lactose, no water.

You should now attempt SAQ 9 on p. 38.

5.6 Summary of Unit 5

In this Unit we selected two exocrine glands, the pancreas and the mammary gland, to serve as examples of exocrine secretion; we considered secretion mainly in cellular terms, giving little space to its physiological control.

Section 5.2.1 was devoted to a brief description of the structure of the gland, and the remainder of 5.2 was concerned with the secretion of protein, that is, the progress of exportable proteins from the ribosomes of the RER to the lumen of the acini. We looked at this largely by following some of the work of Palade and various colleagues, showing how they built up the evidence for the hypotheses they proposed as a result of Palade's earliest work. At first they merely interpreted the EMs as showing the cycle of granule formation, that is, formation in the RER cisternae, followed by the formation of condensing vacuoles and zymogen granules, and then the disappearance of zymogen granules after feeding. They used the existing fractionation techniques to show that fractions from different regions of the cell (i.e. the RER cisternae, condensing vacuoles and zymogen granules) had different proteolytic activity at different stages of the observed cycle.

Later, by using improved fractionation techniques (Fig. 8) and radioactively labelled precursors, they produced more detailed evidence (Table 2) including the involvement of the Golgi complex (though not specifically the stack of Golgi cisternae, sometimes referred to as the 'Golgi body'). By introducing the label into an *in vitro* preparation, they were able to make precise estimates of the time sequences involved (Figs. 10, 11 and 12) by EM autoradiography as well as by fractionation techniques. We gave some consideration to the evidence suggesting that the 'cisternal package' pathway is not necessarily the universal and only one.

Section 5.4 gives a brief account of what is known of the source and mechanism of the secretion of the water and bicarbonate fraction of the pancreatic juice; the debate about the precise role of the pancreatic ducts is touched upon. The Section ends with the general conclusion that the ducts certainly play a large part in the production of this part of the secretion, though whether they modify it by adding bicarbonate or by removing bicarbonate, or possibly by adding it in the fine ducts and tending to re-absorb it (if the flow is slow enough) in the main duct, is still a matter of controversy.

Milk secretion is dealt with in Section 5.5, and the secretion of milk proteins is compared with the secretion of protein by the pancreas. The secretion of fat and lactose do not have counterparts in pancreatic function, and the difference between the fat and protein pathways is underlined. The probable mechanism of lactose secretion is described and this process ties in closely with what is said in the TV programme for this Unit about Golgi function.

Water forms a large part of both pancreatic and mammary secretions, and both are isosmotic with the plasma (or in experimental situations, the perfusate which replaces it). Whereas, in the pancreatic secretion, the principal anion is normally HCO_3^-, and Na^+ is present in a concentration comparable to (though slightly lower than) its concentration in the plasma, in milk from a healthy gland, Na^+ and Cl^- are present in much lower concentrations than in the plasma. In the relevant pancreatic cells Na^+ is probably pumped across the apical surface of the cell into the lumen; in the mammary alveolar cell Na^+ is pumped across the basal surface, back into the plasma. In the pancreas, the large ducts play a part in modifying the secretion; in the mammary gland the ducts are largely impermeable and form a blood/milk barrier, which helps to maintain the consistency of the milk while it is stored in the gland.

The last sentence introduces a further difference between the two glands; secretion in the pancreas is intermittent, depending on stimuli associated with feeding, but rapid when these stimuli are received. Milk secretion, on the other hand, is continuous over a long period of time—from days to years, depending on the species—and the rate of secretion per gram of gland tissue is only a fraction of the capability shown by the pancreas.

Whereas nervous stimulation (via the vagus) and acetylcholine release seems to play a small but significant part in pancreatic secretion, nervous stimulation does not increase milk secretion—its regulation is entirely hormonal.[3] (Stimulation of the sympathetic system can *inhibit* milk secretion, however, by reducing the blood flow to the gland by more than 90 per cent.)

In conclusion, we should remind you that there are other, very important, exocrine glands to be found in species throughout the metazoa, including sweat glands, salivary glands, salt and tear glands and a variety of digestive glands. In ignoring them in this Unit, we are not suggesting that they are less important than those we have considered, or even that they work in exactly the same way. What we are suggesting is that, if you are interested in learning about them, whether by experimentation or by reading the literature (where it exists) you will now know the sort of questions to ask, and which of the answers you get are predictable and which are surprising.

Self-assessment questions

Section 5.2

SAQ 1 (*Objective 3*) In work published in 1956, Palade and Siekevitz attempted to show where the precursors of the pancreatic enzymes were made. (a) What were the general techniques then available and (b) what evidence did they provide?

SAQ 2 (*Objective* 4) Does the evidence presented in the Unit indicate that all the exportable enzyme produced on the RER enters the cisternae?

SAQ 3 (*Objective 4*) What information is gained by comparing the 'enzyme profiles' of the supernatant (cytoplasmic matrix) fractions of homogenates of rabbit acinar cells, before and after stimulation by pancreozymin?

SAQ 4 (*Objective 5*) What evidence, obtained by what technique, led Caro and Palade in 1964 to conclude that the Golgi body was involved (stages 3 and 4(a) in Fig. 13) in the intermediate steps between the RER and the zymogen granules?

SAQ 5 (*Objective 5*) By using the *in vitro* technique of incubating thin slices of pancreas in a controlled medium, Jamieson and Palade (1967a) were able to get much more accurate information on the timing of the intracellular transport steps. Why was this?

Section 5.3

SAQ 6 (*Objective 7(a) and (b)*) Give two pieces of evidence that suggest that in laboratory animals there is a primary pancreatic secretion consisting largely of water and bicarbonate (produced in the acini or intralobular ducts) which may be modified during its passage through the larger (interlobular and main) ducts.

Section 5.5.2

SAQ 7 (*Objective 9*) What are the main items of evidence indicating that the immunoglobulins found in milk are not secreted by the same intracellular pathway as the other proteins, such as casein?

Section 5.5.3

SAQ 8 (*Objective 10(a)*) Are the fats present in milk synthesized in the alveolar cells from precursors, or are they removed from the plasma in the form in which they are secreted, as with the immunoglobulins?

Sections 5.3 and 5.5.5 and the TV programme

SAQ 9 What are the differences (as discussed in the Unit and in the associated TV programme) between the processes leading to the secretion of water in the pancreas and in the mammary gland?

Answers to ITQs

ITQ 1 Any route which is not part of the 'cisternal package' pathway must presumably involve the cytoplasmic matrix. However, the so-called post-microsomal supernatant, that is, fraction 5 of the original 'parent' homogenate, the fraction representing the cytoplasmic matrix, contained little proteolytic enzyme; not more than 13 per cent of the total.

ITQ 2 One such modification would be to try to set up the pancreas *in vitro*, rather than inject the whole animal with the tracer. This would mean that the experimenter could control (a) the concentration of tracer presented to the acinal cells and (b) the timing of the exposure.

ITQ 3 Between +17 and +37 minutes (Fig. 12), so that the radioactive label has time to reach a peak in the condensing vacuole, without appearing in significant quantity in the zymogen granules. In fact Jamieson and Palade administered them at +20 minutes.

ITQ 4 The cycloheximide would show whether the manufacture of new protein was necessary to turn a CV into a zymogen granule. As the enzyme itself is already manufactured and present in the CV, it is really a question of whether new membranes or transport proteins are involved.

The other two would show whether or not respiratory energy is required; such energy might be required for the extrusion of electrolytes from the vacuole, for example.

ITQ 5 (a) The fact that secretin stimulates the flow of electrolytes and water, but not enzyme, but that pancreozymin stimulates the flow of enzymes accompanied by very little water or electrolyte.

(b) The fact that tying off the pancreatic duct, causing cellular degeneration, permanently stops enzyme production more quickly than it permanently impairs the ability to secrete water and electrolytes.

(c) The fact that ethionine destroys the acinar cells and stops enzyme secretion, without stopping water and bicarbonate secretion.

ITQ 6 The figures may imply that some zymogen granules must have a much higher proportion of Tg to ChTg than others and that there is selective release, that is, some granules are released more readily than others. The alternative, and perhaps less likely, interpretation is that the profile of the zymogen granules is constant, and their release random, but that large amounts of Tg are being released in other ways, for example, through the cytoplasm.

ITQ 7 Brew (1969) has suggested that the A protein is attached to the Golgi vesicle, and that α-lactalbumin is synthesized in the RER and passed into the vesicle. Glucose and UDP-galactose would diffuse (or be transported) into the vesicle from the cytoplasm, where the lactose would be produced. Alternatively, the reaction might take place on the vesicle membrane, with just the lactose entering the vesicle itself. This would tend to preserve the glucose and UDP-galactose gradient into the vesicle. It would seem likely that the lactose solution would then be discharged with the exportable protein into the alveolar lumen. The rate of lactose formation could thus be regulated very rapidly by the rate of production, in the RER, of new α-lactalbumin.

ITQ 8 There is selective transport of immunoglobulin, and immunoglobulins can be shown *within* the cell during colostrum secretion.

ITQ 9

	Equilibrium potential between ECF and cell/mV
Na$^+$	+30
K$^+$	−63
Cl$^-$	−17

The ions clearly do *not* distribute themselves passively; Na$^+$ must be being pumped out of, and K$^+$ into, the cell. Cl$^-$, on the other hand, is at a higher concentration in the cell than predicted —it must be retained somehow.

ITQ 10 The predicted values are Na$^+$ −44 mV, K$^+$ −43 mV, Cl$^-$ +44 mV. Thus for Na$^+$ and K$^+$ the values indicate that the distribution is a passive one, the observed and calculated potentials being in very close agreement. For Cl$^-$ however, this is not so, the equilibrium potential being the reverse of that expected, the electrical and chemical gradients being in the same direction. Thus, some other factor must be assisting the retention in the cell of Cl$^-$, possibly a pump, or an exchange with HCO$_3^-$ ion.

Answers to SAQs

SAQ 1 (a) They used EM photography to show differences in the ultrastructure of cells from starved and fed animals; they made cell homogenates and fractionated them by centrifugation and they assayed the proteolytic activity of each fraction, after treating the fractions with a small amount of active trypsin to activate any precursors present.

(b) They showed that the granules in the cisternae of the RER were enzyme precursors, and that these granules could be separated from the RER cisternae (i.e. the 'rough microsomes' of the homogenate) merely by rupturing the ER membrane with detergent, indicating they were not bound to the membrane.

They also showed that in starved animals, in which there was an accumulation of zymogen granules, little enzyme was produced by the RER.

SAQ 2 No, it leaves considerable doubt. It does show that the 'cisternal pathway' is followed by most or all of the trypsinogen and chymotrypsinogen in the guinea-pig, and chymotrypsinogen in the rabbit. However, there is some contradictory evidence about the trypsinogen in the rabbit, ribonuclease in the mouse and amylase in the pigeon (pp. 44 and 45).

SAQ 3 It shows that the ratio of trypsinogen to chymotrypsinogen in the supernatant changes following stimulation. This indicates that relatively more trypsinogen than chymotrypsinogen left the supernatant. As the 'profile' of enzymes remained the same in the cisternal granules and zymogen granules, it suggests that there may also be some other pathway of secretion for trypsinogen.

SAQ 4 They showed that the enzyme was contained in small vesicles in the Golgi region, and in 'condensing vacuoles' in, and above, the Golgi complex, as well as in the RER cisternae and the zymogen granules, thus indicating that the Golgi region was the likely intermediary. The principal technique used was EM autoradiography. They injected ^{14}C labelled leucine into the live animal, subsequently killed it and made EM sections of acinar cells. The emissions from the ^{14}C produced darkened grains on the emulsion coating the sections and the location of these grains gave the information.

SAQ 5 Because they could introduce a short pulse of radio-active amino acid into the cells and then immediately chase it with an excess of unlabelled amino acid. This enabled them to follow a 'wave' of the material through the cell, knowing precisely when it was introduced.

SAQ 6 The fact that there is an inverse relationship between the concentration of HCO_3^- ion and the Cl^-, the HCO_3^- rising with increased flow, but that the sum of the ions remains constant, suggests a bicarbonate-rich secretion which may be modified later.

The results of experiments in which the main pancreatic duct of the cat was perfused indicate that just such modification can actually occur in the duct (Fig. 16).

SAQ 7 (i) Labelled precursors of the immunoglobulins have not been found in the RER, vesicles, etc.
(ii) Labelled immunoglobulins injected into the plasma can be found in the extra-cellular spaces of the gland, within the alveolar cells and in the milk, indicating that they can be transported across the cell intact.
(iii) The fact that the transfer of immunoglobulins from the extra-cellular fluid to the milk can be shown to be selective indicates that there are specific carrier mechanisms on the basal membranes (or the lateral ones, below the tight junctions) and possibly also on the apical ones.

SAQ 8 They are synthesized in the RER from precursors; in mice they accumulate in the RER cisternae, but in goats the droplets appear to form both in the cisternae and in the cytoplasm. The droplets in the apex of the cell are not surrounded by a membrane. The precursors come from various sources in the plasma, including free fatty acids, acetate (in ruminants), triglyceride, and from the hydrolysis of whole chylomicrons by the capillary endothelium.

SAQ 9 In both instances the final fluids, digestive juice and milk, are isosmotic with the plasma, but mechanisms for moving water from the plasma into these fluids are probably quite different.

(a) Fluid secretion in the pancreas depends on respiratory energy; respiratory poisons stop it, and removal of glucose from the plasma stops it, but replacement of the glucose by other energy sources such as pyruvate allow it to continue. Fluid secretion in the mammary gland is also ultimately dependent on respiratory energy; however, although the removal of glucose stops water secretion, its replacement by other energy substrates (such as mannose) does *not* allow secretion to continue. Thus the presence of the actual sugar is necessary, and not just as an energy source.

(b) Na^+ and HCO_3^- are actively extruded from the cells into the ducts in the pancreas, and would therefore be expected to draw water from the cell to maintain osmotic equilibrium. This remains true whether you subscribe to the view that the bulk of the secretion takes place from the centro-acinar cells or the cells of the larger ducts. Ouabain stops water secretion.

In the mammary gland it appears that Na^+, and probably HCO_3^-, are not actively extruded over the apical membrane but are in equilibrium between the milk and the cell. Indeed, Na^+ is pumped out of the cell into the plasma. (Ouabain does not stop fluid secretion immediately, it merely changes the composition of the milk so that it contains more Na^+ and less lactose.) Thus the water is carried out by a non-electrolyte, probably the lactose in the Golgi vesicles.

Bibliography and references

1 General

Suggested reading: Review paper on milk secretion:
Linzell, J. L. and Peaker, M. (1971) *Physiol. Revs.*, **51**.

2 References cited in the text

Brew, K. (1960) 'The secretion of lactalbumin into milk', *Nature*, **222**, 671.
Caro, L. G. and Palade, G. E. (1964) *J. Cell Biol.*, **20**, 473.
Case, R. M., Harper, A. A. and Scratcherd, T. (1969) in *Exocrine Glands* (Botelho *et al.* eds.) University of Pennsylvania Press.
Hokin, L. E. (1955) *Biochem. Biophys. Acta*, **18**, 379.
Jamieson, J. D. and Palade, G. E. (1967a) 'Intracellular transport of secretory protein in the pancreatic exocrine cell: I Role of the peripheral elements of the Golgi complex', *J. Cell Biol.*, **34**, 577.
Jamieson, J. D. and Palade, G. E. (1967b) 'Intracellular transport of secretory proteins in the pancreatic exocrine cell: II Transport to condensing vacuoles and zymogen granules', *J. Cell Biol.*, **34**, 597.
Jamieson, J. D. and Palade, G. E. (1971) 'Condensing vacuole conversion and zymogen granule discharge in pancreatic exocrine cells: metabolic studies', *J. Cell Biol.*, **48**, 503.
Kinura, T. (1969) 'An EM study of the mechanism of milk secretion', *J. Japan Obstet. Gynecol. Soc.*, **21**, 301.
Kurosomi, K. Y., Kobayashi, Y. and Baba, N. (1968) 'The fine structure of the mammary gland of lactating rats, with special reference to the apocrine secretion', *Exptl. Cell Res.*, **50**, 177.
Lee, C. S. and Lascelles, A. K. (1969) 'The histological changes in mammary glands of ewes in response to the local allergic response', *Australian J. Exptl. Biol. Med. Sci.*, **74**, 613.
Linzell, J. L. (1968) 'The magnitude and mechanisms of the uptake of milk precursors by the mammary gland', *Proc. Nutr. Soc.*, **27**, 44.
McDowell, G. H. and Lascelles, A. K. (1969) 'Local production of antibody by bovine mammary glands infused with *Salmonella* flagellar antigens', *Australian J. Exptl. Biol. Med. Sci.*, **47**, 669.
Morris, A. J. and Dickman, S. R. (1960) *J. Biol. Chem.*, **234**, 1 404.
Palade, G. E. (1959) in *Subcellular Particles* (Hayashi, T. ed.) Ronald Press.
Palade, G. E. and Siekevitz, P. (1956) *J. Biophys. Biochem. Cytol.*, **2**, 417.
Rothman, S. S. (1969) in *Exocrine Glands* (Botelho *et al.* eds.) University of Pennsylvania Press.
Schoefl and French (1967) *Proc. Roy. Soc. B*, **169**, pp. 153–165.
Schultz, I. (1969) in *Exocrine Glands*, (Botelho *et al.* eds.) University of Pennsylvania Press.
Scratcherd, T. and Case, R. M. (1973) 'The secretion of electrolytes by the pancreas', *Am. J. Clin. Nutrition*, **26**, 326.
Siekevitz, P. and Palade, G. E. (1960) *J. Biophys. Biochem. Cytol.*, **7**, 619.
Wooding, F. B. P., Peaker, M. and Linzell, J. L. (1970) 'Theories of Milk Secretion: evidence from EM studies of milk', *Nature*, **226**, 762.

3 References to material in prerequisite science courses

1 S100, Unit 18
2 S22–, Unit 9.2.1
3 S22–, Unit 9.2.5 and Appendix 1
4 S22–, Unit 4.8
5 S22–, Home Experiment 4 Film Strip 1
6 S22–, Unit 9.3
7 S22–, TV programme for Unit 8
8 S22–, Unit 10 and the S22– set book, *Control of Water Balance by the Kidney*, by D. B. Moffat.

Acknowledgements

Grateful acknowledgement is made to the following sources for material used in this Unit:

Figures 4 and 13 Holt Rinehart and Winston, Inc. for Novikoff and Holtzman, *Cells and Organelles; Figures 8, 9, 10, 11 and 12* The Rockefeller University Press for J. D. Jamieson and G. E. Palade in *J. Cell. Biol.*, **34**, 2, 1967; *Figure 17* The University of Pennsylvania Press for Schultz, *Exocrine Glands*, 1969; *Figure 18* W. B. Saunders Company for Turner, *General Endocrinology*, 4th ed. 1966; *Figure 20* Schoefl and French in *J. Proc. Roy. Soc.*, B, 169, 1968.

The Course team wish to offer special thanks to Dr Maynard Case for Figures 15 and 16; Dr Jim Linzell for Figure 19 and Dr Peter Wooding for Figures 21, 22 and 23.

Unit 6

Muscles and Contractility

Contents

Table A: part 1

Principal Sections of recommended prerequisite Courses on which this Unit depends

Course	Unit and Section No.	Topic
S100*	3 and 4	tension, load, work
	14	ATP
	28	diffraction
S2–1**	1.4.1, 1.5.2, 1.6	protein structure: primary, secondary and tertiary
	3.2, 3.4	ATP as an 'energy currency'

* The Open University (1971) S100 *Science: A Foundation Course*, The Open University Press.

** The Open University (1972) S2–1 *Biochemistry*, The Open University Press.

Table A: part 2

List of scientific terms, concepts and principles used in Unit 6

Developed in this Unit or its set reading	Page No.	Developed in a later Unitt	Unit No.
actin-containing filaments (thin filaments)	9		
actin molecules	21		
actomyosin	36		
alpha protein	19		
ATPase	18	ATPase	TV 8 and 16
axoneme	38		
calcium activation	16		
catch muscles	28		
cilia	38		
constant-volume phenomenon	9		
contracting muscle	11		
cross-bridge (myosin projection)	7		
dynein	39		
flagella (bacterial)	39		
flagella (protozoan)	39		
force–velocity relation (Hill's equation)	12		
glycerinated muscle	17		
heavy meromyosin	20		
isometric tension	12		
isotonic tension	12		
light meromyosin	20		
long-range force balance	16		
Lymm–Taylor scheme	24		
microtubule	38	microtubule	TV 7
M-line (connections)	16		
myosin head	20		
myosin tail (or rod)	20		
myosin-containing filaments (thick filaments)	7		
myosin molecules	7		
obliquely-striated muscle	27		
paramyosin	23		
paramyosin filament	23		
paramyosin smooth muscle	27		
relaxed muscle	11		
sarcolemma (the muscle cell membrane)	15		
sarcoplasm	15		
series elastic component	11		
smooth muscle	5		
striated muscle	5		

Table A: part 2—continued

Developed in this Unit or its set reading	Page No.	Developed in a later Unit	Unit No.
structural polarity	7		
swelling	17		
tetanus	11		
triad	16		
troponin	22		
tropomyosin	22		
twitch	11		
vertebrate muscle	5		
vertebrate smooth muscle	37		
X-ray diffraction	9		
	In set reading Huxley (1958)		
A-band	,,		
actin	,,		
ATP	,,		
cross-bridge	,,		
H-zone	,,		
I-band	,,		
muscle fibre	,,		
myofibril	,,		
myosin	,,		
phosphate groups	,,		
sliding-filament model	,,		
thick filament	,,		
thin filament	,,		
tropomyosin	,,		
Z-band or Z-line	,,		

Objectives for Unit 6

When you have completed this Unit, you should be able to:

1 Demonstrate knowledge of items listed in Table A by:

(a) Defining them in your own words;

(b) Recognizing the best definition of items;

(c) Recognizing attributes of the items from given alternatives.
(most ITQs: SAQs)

2 By selecting from a matrix or other type of multiple-choice system, or by writing brief statements in your own words, or by drawing simple diagrams:

(a) Recognize the essential features of the 'sliding-filament' model for changes in muscle length;
(ITQ 1; SAQs 5 and 17–20)

(b) Given the 'sliding filament' model, recognize features likely to be concerned with force generation and contraction in muscle;
(ITQ 2; SAQs 5 and 17–20)

(c) Describe the general structure of the thin and thick filaments in striated muscle;
(SAQs 1 and 2)

(d) Recognize the difference between twitch tension and tetanic tension;
(SAQ 3)

(e) Show how the length–tension diagram is related to the 'sliding-filament' model;
(SAQ 4)

(f) Recognize the function of the major components in the sarcoplasmic reticulum;
(SAQs 6 and 7)

(g) Distinguish between the situations and functions of the proteins actin, myosin (and its two subfragments), paramyosin, tropomyosin and troponin
(SAQs 9–16)

(h) Place the site of the muscle ATPase in the protein system.
(SAQ 10)

3 By selecting from a matrix or other type of multiple choice system, or by writing brief statements in your own words, or by drawing simple diagrams:

(a) Describe the structure of obliquely-striated muscle, paramyosin smooth muscle, vertebrate smooth muscle, cilia, microtubules and bacterial flagella;
(SAQs 21–28)

(b) Describe the basic structure of various filaments in the systems mentioned in 3(a) above;
(SAQs 22, 23, 25, 28 and 29)

(c) Recognize the different functions of sub-structures in the systems mentioned in 3(a) above;
(SAQs 21 and 31–33)

(d) Place the site of the ATPase (if any) in the systems mentioned in 3(a) above;
(SAQ 30)

(e) Recognize and offer physiological explanations (or possible explanations) for the varieties of system mentioned in 3(a) above.
(SAQs 24, 26, 27 and 31–33)

4 Orally, or by writing a short essay, discuss current ideas of mechano-chemical conversion in contractile systems.

6.0 Introduction and study guide to Unit 6

6.0.1 The plan of Unit 6: an overall study guide

This Unit represents more than one week's work, and should probably take you about one and a half times as long as an average Unit (we have, of course, made allowance for this in the overall scheme of S321).

Unit 6 divides naturally into two parts, which probably will take you roughly equal amounts of time. These are the parts up to and including Section 6.4.6 and the subsequent part. To punctuate this natural break we have included a series of SAQs (1–20) on pp. 25 and 26.

We begin the Unit by considering striated muscle (Sections 6.2–6.3). First we describe its structure, in Section 6.1, where you are required to read H. E. Huxley's article (Chapter 5) in the *Biological Bases of Behaviour**. A postscript to Huxley's article follows and we widen the discussion to include some of the physiology of striated muscle and the coupling between the contractile impulse and the mechanical response (Section 6.2). Then we go on to consider the physical stability of striated muscle (Section 6.3).

After a consideration of the muscle proteins in Sections 6.3.1–6.4.6, the series of SAQs (1–20) picks out the essential features in the first half of the Unit.

We go on, in Sections 6.5–6.5.4, to describe other types of muscle, stressing their similarities and dissimilarities to striated muscle. In Sections 6.6–6.6.2 we widen the description to include non-muscular contractile systems, including micro-tubules, cilia and bacterial flagella. A futher series of SAQs (21–34) highlights the essential features in these Sections.

Finally, in Sections 6.7 and 6.8 we consider the characteristics of the basic contractile event, and describe some current speculations about the form of this event.

There are three papers (Ebashi 1972, Elliott 1973 and Huxley 1969**) associated with this Unit. You are not obliged to study them as they are black-page material; but they could usefully be read where indicated in the text.

6.0.2 Introduction to Unit 6

One of the aims of a modern biologist is to link the molecular structure of living tissues with the function of those tissues in the organism. Muscle is an attractive tissue to a biologist, particularly a physically-minded biologist, for three reasons: first, because it has regularities at microscopic and sub-microscopic levels which may be investigated by suitable physical techniques; secondly, because muscles may be isolated and their working properties determined by the methods of quantitative physiology; and thirdly, but most important, because muscle is a biological machine for transforming chemical energy into mechanical work. Muscle has been much studied by many techniques since the pioneer days but, as we shall see in this Unit, the physical and chemical processes of muscular contraction on a molecular level remain controversial. Why is this controversy so fascinating? Because the final goal is to see an enzyme in action—at the cellular level—converting the metabolic energy of ATP into the mechanical work of muscle contraction. Indeed, it is precisely because muscles are so regular and uniform, and open to so many different techniques, that they offer the best chance of achieving this goal, which will surely have relevance to all the other ATPase systems.

When we observe an animal, we usually observe its reactions to situations. These reactions sometimes involve visceral functions, such as the secretion of saliva, but the majority include some form of muscular activity. A complete description of the physiology of animals should therefore include a study of the muscular

---

* Open University set book for SDT 286. Ed. Chalmers Crawley & Rose. Publishers, Harper and Row (1971). This article is also available as *Scientific American* Offprint No. 19. Price 11p (cash with order) by post from W. H. Freeman and Co., 58 Kings Road, Reading, England RG1 3AA.

** These papers are part of the offprint collection supplied to Open University students with the supplementary material. Full titles are given in the Bibliography and References on p. 53.

system and of the properties of muscle cells. There are many differences between the various phyla of animals in the types and arrangement of the muscles or other contractile systems. At the same time, certain principles emerge from a general study of contractile systems, and by the end of this Unit we should be able to see these principles.

Striped or 'striated' muscles, which we shall describe in Sections 6.1–6.4.6 are so called because, seen in the light microscope, they have obvious transverse stripes. They make up the bulk of vertebrate muscles. Typically, these are 'voluntary', that is, they can be consciously controlled in man, and are innervated through the somatic motor nerve components. Another type of striated muscle is present in the vertebrate body—'cardiac' muscle, which makes up the bulk of the heart. Some of its properties were discussed in S100.[1] Cardiac muscle is classified as striated muscle on structural grounds, because it is striped, but it has several modifications from the point of view of innervation and control. However, we do not have the space here to consider these modifications.

striated muscle

vertebrate muscle

As we shall see, striated muscle is not confined to the vertebrates. Historically, however, a considerable proportion of the experimental work on muscle has used vertebrate striated muscle. Over the past 20 years a large body of evidence has grown up to support the 'sliding-filament' model of striated muscle. This model was originally proposed in 1953, at the same time but independently, by two groups of workers, A. F. Huxley and R. Niedergerke in Cambridge, England and H. E. Huxley and J. Hanson, English scientists who had been working together temporarily in Cambridge, Massachusetts. We shall begin in Sections 6.1–6.4.6 by describing the structure, and some of the physiology, of vertebrate striated muscle, and the major properties of the contractile proteins. In the following part of the text we shall widen the scope to deal with the several types of muscle which are not striped or striated—or at any rate not striped in the conventional sense of the term. One of these is the smooth muscle of vertebrates, which occurs in the viscera and the walls of the blood vessels. In vertebrates, smooth muscles are typically not under voluntary control, they are innervated through the sympathetic and parasympathetic nerves (the autonomic nervous system[2]).

smooth muscle

Leaving aside the innervation, there are two major aspects of a muscle fibre. The first aspect is the architecture of the fibre itself—whether it is striated or smooth and how the various protein components are arranged within the fibre. The second aspect is how, in precise physical–chemical terms, the fibre works—how the chemical energy is transduced into mechanical work by the mediation of the molecular structure.

Over the last two decades a very great deal has been learnt about the architecture of the fibre and we shall start by describing the current state of knowledge for vertebrate striated muscle. The second aspect is still controversial, and we shall discuss at least two viewpoints at the end of this Unit.

6.1   The structure of vertebrate striated muscle: the sliding filament model

As we have already mentioned, H. E. Huxley was a founding father of the 'sliding-filament' model for striated muscle. In 1958 he wrote an account of this model for *Scientific American*. *This article is reprinted as Chapter 5 of The Biological Bases of Behaviour (about 3 500 words) which you should now read.*

> **Study Comment** You are studying Huxley's article to achieve Objective 2(a) and (to a lesser extent) 2(b). Start by re-reading these objectives, and keep them in mind while you study the article. Remember that inevitably, when you meet a piece of scientific literature, you will find things in it—details or digressions— which are not relevant to your own purposes. You must develop the courage to go on past these, and not be hung up on them. In this way you will learn to extract the bones of someone's contribution, and only if necessary will you have to spend time examining the flesh, limb by limb. (You may *need* to do so if you doubt the major conclusions; remember that scientists, even distinguished ones, are sometimes wrong, at least in emphasis.)

In Huxley's article the first page, and some of the second page, contains information which you either know already, or which is irrelevant to our objectives. *Do not waste time on it*—though by all means skim though. You may find it triggers your interest in things (such as the energy budget) which are outside the scope of our present objectives. If it does, and if you follow this interest up in some of the bibliography items (p. 53), you are on the way to your own participation in a scientific field.

For the purposes of Objectives 2(a) and 2(b) the important text begins on p. 71, third paragraph, 'from the chemical point of view. . . .'

Notes on terminology

Surely no one who has read S100 will need to be told that 'tension' is another word for the force[3] produced by a muscle. In SI units the unit of force is the newton, so the tension of a muscle is expressed as newtons per square metre of cross-sectional area. A mammalian muscle can exert a maximum tension of about $4 \times 10^5 \, \mathrm{N \, m^{-2}}$ at 37 °C.

The 'load' on a muscle is the external force on it (e.g. a weight which it must lift; see Fig. 9 on p. 12). In order to shorten, the muscle must produce enough force to overcome this load.

When the muscle shortens (moves) and moves the load, it does 'work'[3]. The work done on the *external* load is sometimes called 'external work'. Work done by the muscle in shortening against its own *internal* (i.e. viscous[4]) resistance or friction is sometimes called 'internal work'.

Note for students who have completed SDT 286

If you studied Huxley's article when you read SDT 286, and if you can go back to the appropriate SAQs in SDT 286, Unit 4 (first year only) and answer them correctly, you need not read it again now. Proceed with the next Section (6.1.1).

6.1.1   Postscript to Huxley

The article you have just read was written more than 15 years ago. Since then further research has filled in, and developed, many details of the structure of striated muscle, but has not altered the basic picture described by Huxley. Indeed, the sliding-filament model is now taken as axiomatic in all discussions of striated muscle. Refinements in electron microscopy have shown that the myosin filaments (thick filament), about 15–20 nm in diameter in frog or rabbit muscle, is made up of 'tadpole-like' molecules, doubled-headed tadpoles in fact (see Fig. 1(a) and (b)). The direction of these molecules in the filament reverses at the centre of the thick filament (see Fig. 2). The heads of these tadpole-like molecules form the 'cross-bridges' which reach out towards the actin filaments. Because the direction of the molecules is reversed at the centre of the filament, the whole filament can be considered as having a *structural polarity* towards the

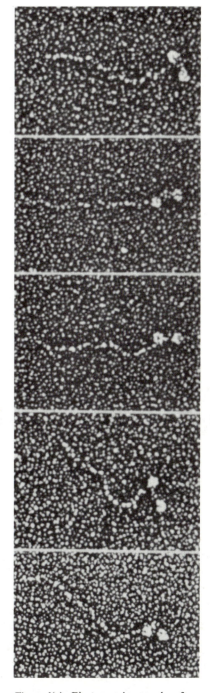

Figure 1(a) Electron micrographs of myosin molecules, made by the shadowing technique (Unit 1), showing the double-headed molecules. ($\times 24\,000$)

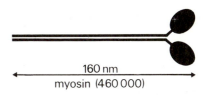

160 nm

myosin (460 000)

(b) A diagram of the double-headed myosin molecule. The total molecular weight is about 460 000. The two-chain tail is shown separated, but is in fact twisted into a two-strand rope.

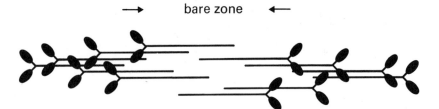
bare zone

Figure 2 Diagram showing how the first few double-headed molecules pack tail to tail in the myosin filament, and how subsequent molecules add on in the two directions to give a bipolar filament (the directions indicated by the arrows) and a central bare zone.

filament centre (see the arrows in Fig. 2). This polarity is symmetric about the filament centre. Another consequence is the bare zone at the centre of the filament, where there are no 'cross-bridges'.

(a)   (b)

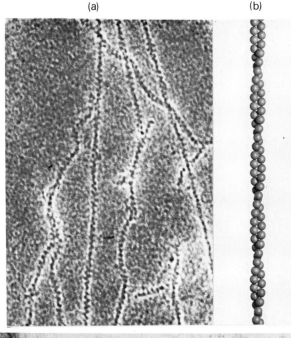

Figure 3(a) Electron micrograph of actin filaments, made by the negative-staining technique (Unit 1), showing the characteristic beaded appearance. (× 34 000)

(b) A model made with strings of table-tennis balls, to show the two strings of molecules loosely coiled about each other.

Figure 4(a) (*left*) An I-band segment, with actin filaments attached to both sides of part of the Z-line structure. (× 65 000)

(b) (*right*) A similar segment treated with myosin heads to show the arrow-head structures. Notice that on each side the arrow-heads point away from the Z-line remnants. (× 65 000)

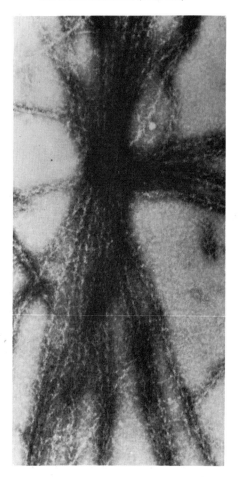

The actin-containing filaments (thin filaments), which are about 8–10 nm in diameter in frog or rabbit muscle, themselves have a structure like two strings of beads wound loosely together (see Fig. 3). These thin filaments also have a structural polarity and in this case the polarity reverses at the Z-line. This has been shown by an elegant technique in which the thin filaments are decorated with myosin fragments (see Fig. 4). Further details of these experiments are given in a subsequent article written by Huxley (1965) for *Scientific American*. If you think you can spare the time, the second article is well worth reading, but we have not prescribed it because it will most likely take you beyond your time allowance.

Another structural feature of striated muscle is that the assembly of filaments works as a constant-volume system. This is true both when the muscle is at rest and during contraction. If S is the sarcomere length (the length between successive Z-lines) and a is the centre-to-centre distance between myosin filaments, then Sa^2, which is proportional to the volume occupied by each myosin filament, is found by experiment to be a constant (see Fig. 5(a)). This experiment can be

thin filament

constant-volume phenomenon

Figure 5(a) The inverse-square of the interfilament separation plotted against the muscle length for resting striated muscle (filled points) and for contracting muscle (open points). A straight line passing through the origin shows that the muscle works at constant volume (Sd^2 = constant). In this case the overall length of the muscle was measured rather than the sarcomere length in the operative region, and because the ends of the muscle are less stretched than the middle the plot does not pass precisely through the origin.

carried out with reasonable accuracy, using light diffraction[6] to measure S (which in frog muscle is between 2.0 μm and 3.6 μm in the intact animal) and X-ray diffraction[6] (see also Appendix 1) to measure a (which is typically 43–32 nm in the same muscle), both on the intact living muscle fibre. To make the X-ray measurement on a *contracting* muscle a stroboscopic technique[7] must be used. This is because the X-ray technique needs exposures of the order of minutes or tens of minutes, and a muscle fibre can be made to produce tension for only a second or so at a time and then needs a rest of 15 seconds or so before it is stimulated again. X-ray patterns from contracting muscle were first taken in 1964, by Elliott, Lowy and Millman in London, and Figure 5 is adapted from their publications. To a first approximation the lattice volume is the same in resting and contracting muscles.

The implications of this work for models of muscle contraction are interesting. On the assumption of an effective radius of 5 nm for actin and 10 nm for myosin, the surface-to-surface distance between actin and myosin filaments varies between

X-ray diffraction patterns

about 4 nm and 10 nm—a factor of 2.5—over the range of lengths at which muscle contracts in the body (see Fig. 5(b)). As the contractile event involves some unspecified interaction between the two proteins, as you read in the Huxley article, the interaction must be capable of taking place in this variable environment. To put it in another way, the cross-bridges must be able to accommodate to the changing surface-to-surface distance.

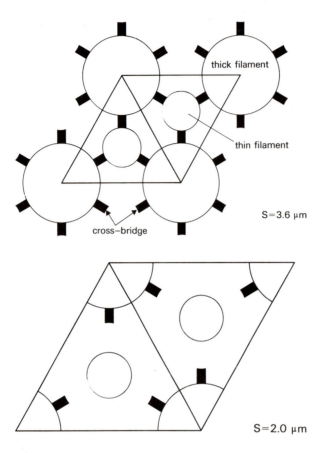

Figure 5(b) The relationship of thick and thin filaments in (frog) muscle at sarcomere lengths of 2.0 μm and 3.6 μm. The myosin projections are shown just touching the thin filaments at a sarcomere length of 3.6 μm. If they do not change in length they will fail to bridge the gap at a sarcomere length of 2.0 μm, unless they are connected flexibly, rather than rigidly, to the filament backbone.

At this point we can mention two more results from the X-ray experiments on contracting muscle. We do not need to explain details of the interpretation of X-ray photographs—this is a highly technical matter, though if you are interested you can get some idea from the S100, TV programme 28 and from Appendix 1 of this Unit. For our purpose all you need to know is that suitable regularities in the structure give rise to recognizable, and measurable, features in the X-ray diffraction photographs. In muscle such features are the distances of the molecular repeating arrangements on the myosin and actin filaments, and the regular arrangement of the cross-bridges.

The X-ray experiments established that when striated muscle contracts (and changes length by 10 per cent or more):

1   The molecular repeat pattern, of the actin along the thin filaments and the myosin along the thick, stays the same to within 1 per cent—so these filaments do not shorten.

2   The pattern arising from the regular cross-bridge arrangement almost completely disappears so that the cross-bridges themselves are taking part in the contractile process; they are displaced from their resting positions and are probably in asynchronous cyclic movement.

These experiments were carried out by H. E. Huxley and his colleagues at Cambridge, as well as by the London group we have already mentioned.

6.1.2   The physiology and mechanical properties of vertebrate striated muscle

Probably you have already met 'antagonistic' sets of muscles in invertebrates: for instance, the circular and longitudinal muscles of one segment of an earthworm. Vertebrates differ from those invertebrates you have studied in having an internal, jointed skeleton. Some of the joints between bones allow movement in one plane only (as your knee flexes and extends but scarcely bends in any other way); other joints (such as that at the base of your thumb or at your shoulder) allow movement in many directions. For the knee joint, the antagonistic muscles are the flexors and extensors*.

Note, that, as in all muscles, the active change in a vertebrate muscle is shortening. Stimulation of a muscle, either directly or through its nerve supply, leads to contraction. When one muscle contracts, the antagonist, if it is relaxed, is stretched passively; it is then itself ready to contract and when it does this, it will stretch the first muscle which is probably now in a relaxed state. Of course, both muscles of an antagonistic pair may be active simultaneously and when this happens the position of the joint will depend on the balance between them— you can hold your knee joint in any position you choose between full flexion and full extension.

Largely because of the pioneer work of A. V. Hill and his school there is a wealth of information about the mechanical properties of striated muscle. We do not have the space to go into this in detail here, and shall merely pick out some of the salient points. If this whets your appetite, you can go on to read Chapters 6–8 in Bendall's book (see the Bibliography on page 53).

contracting muscle

relaxed muscle

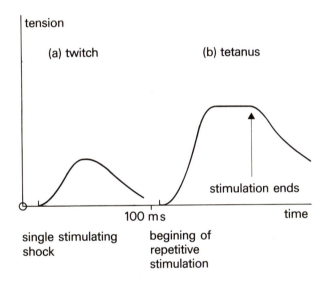

The response of a muscle fibre to a stimulating shock takes the form of a 'single twitch' (see Fig. 6). This is an all-or-none response (cf. the nerve action potential), that is, either there is a response or there is not, and if there is the amount of tension does not depend on the strength of the stimulus. If the stimulus is repetitive, and the repeat rate is too fast to allow single twitches, the responses fuse into a *tetanus* (see Fig. 6). The peak tension in a single twitch is less than in a tetanus and this led Hill to suggest that inside the muscle there is an elastic part which he called the *series elastic component*, in series with the contraction-producing part. Work must be performed by the contractile apparatus in stretching this series elastic component (see Fig. 7). In a single twitch, Hill hypothesized that the elastic component is not fully stretched and thus maximum tension is not achieved. Once fully stretched, as in *tetanus*, the elastic component remains so

Figure 6 Diagrammatic tension–time curves for a typical striated muscle. A single twitch, in response to a single stimulus is shown as (*a*) and a tetanus, in response to repetitive stimulation is shown as (*b*).

twitch

tetanus

series elastic component

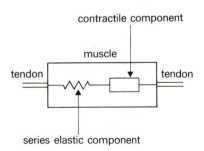

Figure 7 A. V. Hill's hypothesis. The muscle contains a contractile component, and a series elastic component which must first be stretched when the contraction is initiated.

---

* If you read anatomical and medical textbooks you will find very many terms used to describe the functions of muscles. These terms are not important from our point of view and you are not required to know any of them. They describe the types of movement made by the limb or other organ which the muscle moves. For example, some muscles cause rotation about the long axis of a limb or digit. In the special case of the forearm, these rotatory movements are called 'pronation' and 'supination' (turning with the hand down and up respectively). Movement at a joint between body and limb is described by assuming that the vertebral column is horizontal and parallel with the ground—this is the normal posture of vertebrates. Muscles that move the limb towards the head are called 'protractors'; their antagonists that pull the limb backwards are called 'retractors'. If you kneel in front of a chair and bend your chest over the seat you can work out these movements for your arm in relation to the shoulder joint. Because the normal human posture is with the vertebral column vertical instead of horizontal, protraction of the forearm when you are standing up means lifting the arm so that it points vertically upwards. Muscles may have mixed functions: extensor-rotators, flexor (of one joint) -protractors (of another joint) and so on.

and no extra work is required. The hypothesis is supported by the fact that if a muscle is stretched quickly at the onset of a twitch (by an amount necessary to take up the slack in the elastic component), then the tension produced is equal to that produced in tetanus. Once stretched, the series elastic component, as well as the contractile apparatus, will be capable of producing tension. The concept therefore implies that the tension recorded externally will endure beyond the point at which the contractile part ceases to produce tension actively, that is, it implies that tension endures beyond the 'active state'.

To show whether a muscle is (or is not) in an active state, a quick-release method is used. The muscle is made to contract under conditions where it is not able to shorten. Then the muscle is allowed to shorten by a known small amount, and this causes the tension to drop suddenly. If the muscle is 'active'

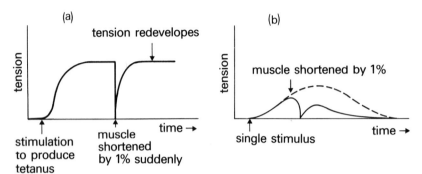

Figure 8(a) A 'quick-release' experiment during a muscle tetanus. The muscle is totally active, and the whole of the tension is redeveloped after 1 per cent shortening.

(*b*) When the same experiment is tried during a single twitch, only part of the tension is redeveloped.

the tension will redevelop at the new, shorter length. As might be expected, tension is fully developed if the shortening occurs during tetanus (see Fig. 8(a)). However, if shortening occurs during a single twitch, little if any tension redevelops unless shortening occurs very early in the twitch. This latter result (see Fig. 8(b)) indicates that the active state has begun to decline well before maximum external tension is reached in the twitch. From recent studies it is apparent that the active state is associated with the presence of free calcium ions inside the muscle fibre (see Section 6.2). The concept of active state is of importance in some types of smooth muscle.

6.1.3 Isometric tension and isotonic tension

So far we have been talking implicitly about *isometric tension*, that is, the tension that a muscle develops when it is stimulated but its length is held constant. Effectively, we may think of the muscle as working against an load equal to or higher than the isometric tension; therefore it cannot shorten. But suppose that the load is less than the isometric tension and that the load is a given weight P (see Fig. 9). Then the tension produced is the *isotonic tension* to overcome P. If the muscle can overcome P, and lift the load or weight, it will do so with a certain velocity, which we call the 'shortening velocity'. Hill showed that this velocity, v, is related to the load P by the following equation:

$$(P + a)v = b(P_0 - P) \tag{1}$$

P_0 is the isometric tension that the muscle can develop at the length under consideration; a and b are constants for the muscle. This looks a complicated equation, but expresses a very simple fact: when the muscle is loaded more heavily it shortens more slowly; when it is loaded with a weight equal to (or in excess of) the maximum isometric tension, it does not shorten at all (see Fig. 10).

6.1.4 The length–tension diagram

The isometric tension which a muscle can produce depends upon the length of the muscle. These days we can measure the sarcomere length S simply by light diffraction (as you have seen) so that the muscle length is conveniently expressed in terms of S, rather than as the overall length of the muscle. (See Fig. 5.2 in Huxley's article for the relationship between the two lengths, remembering that the sarcomere length S is the distance between adjacent Z-lines.)

To remove ambiguities in early experiments where tension was recorded as a function of length, A. F. Huxley and his colleagues at University College, London

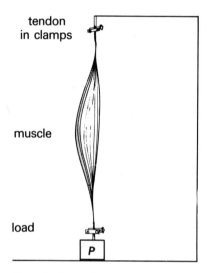

Figure 9 A muscle working against a load. If the weight is too large for the muscle to move, it is said to produce tension 'isometrically'. If it can lift the load, the tension is 'isotonic', the load will be lifted at some finite 'shortening velocity'.

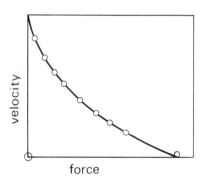

Figure 10 Hill's force–velocity curve for a typical striated muscle, showing how the velocity of shortening depends on the load lifted.

have developed a most ingenious experiment in which the sarcomere length is held constant by an optical and mechanical feedback system. The isometric tension is then recorded at a precisely determined length. The experiments were done on single muscle fibres, to avoid problems from connective tissue. The results of these experiments are shown in Figure 11. The correspondence between the length–tension diagram and the amount of overlap of the filaments is very interesting and gives much information about the contractile process.

Figure 11(a) gives the 'standard' lengths (that is, the generally accepted lengths) of the filaments in frog muscle. The thick filaments (a) are 1.6 μm long, and have a bare zone (c) of about 0.15–0.2 μm. The thin filaments (b) are 2.05 μm long, and the Z-line thickness (z) is about 0.05 μm. (Note that in fact there is a discontinuity in the thin filaments at the Z-line. This detail does not affect the

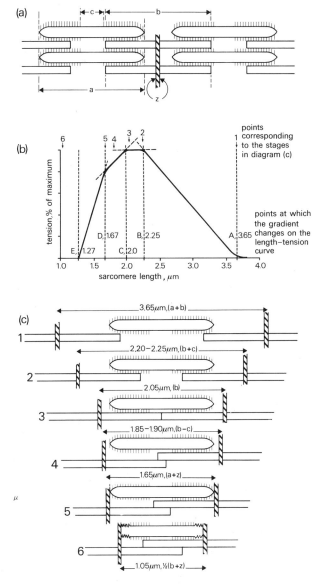

Figure 11(a) Standard filament lengths. a = 1.6 μm; b = 2.05 μm; c = 0.15–0.2 μm; z = 0.05 μm.

(*b*) Tension–length curve from part of a single muscle fibre (schematic summary of results). The arrows along the top show the various critical stages of overlap that are portrayed in (*c*).

(*c*) Critical stages in the increase of overlap between thick and thin filaments as a sarcomere shortens.

Note In studying this Figure, notice that in (b) A–E are experimentally-determined points at which the gradient of the length–tension curve changes. Points 1–6 in (b) are theoretical points predicted on the sliding-filament model as shown in (c).

argument given here, though.) In Figure 11(c) various sarcomere lengths are shown, numbered 1–6. These are the critical stages in the increase of overlap between thick and thin filaments as the sarcomere shortens. 1 is the point of zero overlap, at a sarcomere length of 3.65 μm. At sarcomere lengths longer than this there is no tension (or very little, which is another story) because there is no overlap. Between 1 and 2 the number of cross-bridges which are overlapped increases linearly with the decrease in the length of the sarcomere, and the tension produced increases linearly also, as is shown in Figure 11(b). At point 2 (sarcomere length, 2.20 μm) the ends of the actin enter the bare zone, and a further decrease in sarcomere length does not cause a further increase in the number of cross-bridges that are overlapped. Also the tension produced stays constant, as shown in Figure 11(b). At point 3 (sarcomere length, 2.05 μm), the ends of the thin filaments butt against each other, and from 3 onwards there is 'double overlap' of the thin filaments. The tension produced now falls off with further decrease of the sarcomere length. As you see in Figure 11(b) the rate of fall-off of tension increases at a sarcomere length of 1.67 μm, which is close to the sarcomere length at which the thick myosin filaments butt against the Z-line (point 5). At a sarcomere length of 1.27 μm the tension has decreased to close to zero. These observations agree well with the hypothesis that the tension produced is a sum of 'quanta' of tension generated at each overlapped 'cross-bridge'. This hypothesis immediately explains the initial linear increase of tension. While the ends of the actin filaments remain in the bare zone, but do not overlap each other, there is no increase of overlapped cross-bridges and therefore there is no further increase in tension (points B–C in Fig. 11(b)). After point C, the tension falls off, perhaps initially because the double-overlapped actin filaments interfere with each other, and then also because the portions of the actin filaments in the reverse half of the sarcomere actually produce negative amounts of tension, because of their directional properties. (Remember that both actin and myosin have polar structures.) After point D in Figure 11(b) the tension falls off still further, because the myosin filaments now butt against the Z-lines, and presumably absorb part of the force which is produced. In addition, the filaments will become crumpled, and this will interfere with, and decrease, the number of active cross-bridges.

Now attempt ITQs 1 and 2 and then read the answers and comments on p. 51.

ITQ 1 (*Objective 2(a)*) Which items in the matrix are *essential* elements of a sliding-filament hypothesis for changes in length in muscle or other contractile systems?

(*Note* As far as we know, all the statements made in the matrix items are true.)

(i) Actin and myosin can be extracted differentially with appropriate changes in the band densities.	(ii) The myosin molecules are double-headed.	(iii) Both the thick and thin filaments have polar structures	(iv) The thick and thin filaments move relative to each other as the length changes.	(v) The faster the muscle changes its length during contraction, the less tension it develops.
(vi) Both thick and thin filaments are discontinuous along the fibre.	(vii) Myosin contains an ATPase enzyme, situated in the heads of the molecules.	(viii) Actin and myosin are segregated in separate thin and thick filaments.	(ix) Under certain circumstances cross-bridges connect the thick and thin filaments.	(x) Suitably treated muscle will contract when ATP is added, and show changes in band pattern.
(xi) The contractile mechanism maintains constant volume when the muscle changes length.	(xii) There is no appreciable change in length of either type of filament during changes in muscle length.	(xiii) The shape of the length–tension diagram.	(xiv) Hill's equation.	(xv) Tension recorded externally endures beyond the active state.

ITQ 2 (*Objective 2(b)*) (a) Assuming that a sliding-filament model is defined, which items from the matrix would you choose if developing possible mechanisms of force-generation within the model?

(b) Which items would you expect to arise as consequences of the mechanism which you develop?

ITQs 1 and 2 attempt to differentiate between structural features of a sliding-filament mechanism, and the features that relate to the development of tension within the architecture defined by the sliding filaments. This leads us to the most important question: how is the tension developed?

This question has no agreed answer; it is an area of current scientific controversy. The question will be discussed at the end of the Unit in Section 6.8.

6.2 The sarcoplasmic reticulum and the contractile impulse

The sarcoplasmic reticulum is a system of membrane-bound vesicles, which corresponds to the endoplasmic reticulum in other cells (see Unit 1).

In striated muscle cells the sarcoplasmic reticulum is generally rather regularly arranged, though the details of the arrangement vary from animal to animal, and sometimes from muscle to muscle within the same animal, depending on the physiological function of the muscle (e.g. whether it is specialized for slow or fast contraction). In smooth muscle cells the sarcoplasmic reticulum is not generally so well developed, and less is known about the detailed arrangements.

sarcoplasm and sarcoplasmic reticulum

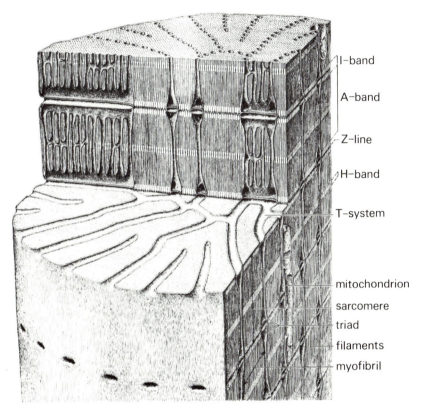

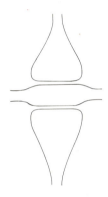

Figure 12(a) (*left*) The sarcoplasmic reticulum in frog muscle, and the relationship to the transverse tubules and the myofibrils.

(*b*) (*above*) shows the triad structure on an enlarged scale.

The relation between the sarcoplasmic reticulum and the muscle membrane shows essential features which are the same for all striated muscles; these are shown in Figure 12(a) in the configuration found in frog muscle. The first essential feature is a transverse invagination of the cell membrane, which is continuous with the extracellular space and extends across the cell. In close opposition to this is the second essential feature, longitudinal elements of the

cell membrane (or sarcolemma)

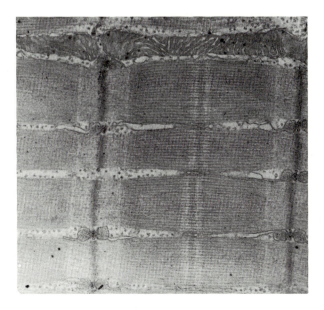

Figure 13 Electron micrograph of a frog muscle fibre which has been bathed in iron-bearing ferritin. The ferritin has here penetrated only the central transverse element of the triads and the longitudinal elements are free of ferritin. Notice that the triad structure in this muscle is precisely next to the Z-line. (25 000)

sarcoplasmic reticulum which extend parallel to the myofibrils. The juxtaposition of these two elements gives the *triad* system, which is shown diagramatically in Figure 12(b). 'Triad' refers to the sandwich of three elements.

triad

One of the differences between types of striated muscle is the position of these triads. In rabbit muscle they are at the edge of the A-band, but in frog, and many other cold-blooded species, they are by the Z-lines as shown in Figures 13 and 14. The transverse tubular system has been perfused, from outside the fibre, with the iron-bearing protein ferritin, which is visible as small dark granules in the micrographs (of thin sections) because its high iron content makes it scatter electrons. This technique makes clear the continuity between the transverse tubular system and the extracellular space. Note that the ferritin has not penetrated into the longitudinal elements of the sarcoplasmic reticulum, which are truly intracellular.

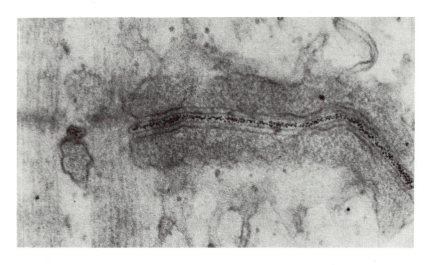

Figure 14 A longer run of ferritin-bearing transverse tubule, beside the Z-disc. The neighbouring vesicles are free of ferritin. ($\times 75\,000$)

It seems that when the cell membrane of a muscle fibre is depolarized (just like a nerve[8]), the depolarization is conducted inwards along the transverse system. This effects the longitudinal system through the close contact at the triad (the precise mechanism is unknown). The longitudinal system then releases Ca^{++} ions into the fibre and these ions react with the contractile proteins and initiate contraction (see Section 6.3.4).

calcium activation

When the depolarization has passed, the longitudinal elements of the sarcoplasmic reticulum actively sequester Ca^{++} again, and the contraction ends (cf. Units 3 and 4).

6.3 The stability of the muscle lattice and the long-range force balance

A striking feature of striated muscle, and of many other muscles, is the stability and precision of the arrangement of the lattice of filaments—all the X-ray and electron microscopy evidence testifies to this regularity. An immediate question arises: what properties of the filament assembly give rise to such stability? In a muscle at rest there can be no large contribution from the cross-bridges because these are conventionally not supposed to be attached in the resting state. There are, it is true, a few permanent cross connections, which presumably are elastic, between the myosin filaments at the centre of the A-band. These are called M-line connections, but they in themselves do not seem sufficient to account for the stability of the muscle lattice, and for some striated muscles they seem not to occur at all.

long-range force balance

M-line connections

It seems almost certain that the over-riding feature that makes for stability is the long-range interactions between the muscle filaments themselves. These interactions are of two types:

1 All matter interacts with other matter in a quantum-mechanical way, giving rise to the forces that are called van der Waals or dispersion forces. Van der Waals forces in general are attractive forces, and each filament is attracted to all its neighbours.

2   The van der Waals attraction is balanced by a repulsion which arises because all the filaments in a muscle are negatively charged. The charge on a protein arises because many of the amino-acid residues are ionized at physiological pH. The basic groups carry a positive charge and the acidic groups a negative. Usually the acidic groups are greater in number and the net charge is negative.

In fact the muscle filaments are quite strongly negatively charged; the charge on a thick myosin-containing filament is probably of the order of 40 000 times the electronic charge and the charge on a thin actin-containing filament is not likely to be very different. These negative charges, together with the ionic nature of the interfilament medium (containing about 100 mM of the K^+ ion), give rise to forces, which are called 'electrical double-layer forces'. The electrical double-layer forces are repulsive; they tend to hold the filaments apart. It is, then, the balance between these attractive and repulsive forces that holds the muscle lattice in its stable position, and overcomes the disruptive effect of Brownian motion.

It is possible to show the nature of these forces by using a preparation first developed by Albert Szent-Györgyi. This preparation, called a glycerinated muscle, is a muscle where the membrane has been destroyed by treatment with glycerol. The contractile mechanism is preserved essentially intact so that the medium round the filaments can be changed by simply diffusing in a different medium. The inter-filament spacing at a constant muscle length (or sarcomere length) can then be followed using the X-ray diffraction technique and Figure 15 shows typical results, taken from the work of Elizabeth Rome. The effect of a change of ionic strength, or of pH, of the inter-filament medium is to cause the lattice to swell and shrink; the total change can be as much as 30 per cent.

glycerinated muscle

swelling

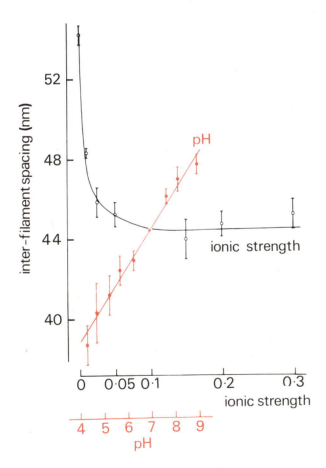

Figure 15 The centre-to-centre distance between myosin filaments in the A-band of glycerinated rabbit psoas muscle, as the pH and ionic strength of the bathing medium is changed. Note that this behaviour depends on the sarcomere length, here 2.9 μm.

These pH and ionic strength effects have been understood in principle for a long time in systems of charged particles suspended in ionic solutions (colloidal systems). As the concentration of the ionic solution is increased, the electrostatic repulsion between the particles decreases and the particles move closer together. (This is called the screening effect.) As the pH is made increasingly alkaline, the charge on the particles becomes increasingly negative as more of the amino-acid residues are ionized on the surfaces; the electrostatic repulsion increases and the particles move further apart. A further parameter can be shown to affect the inter-filament spacing of muscles in these experiments. This is the valency of the positive ion. If the medium is changed from a solution containing the univalent potassium ion to one of similar ionic strength but containing a divalent positive ion, calcium for example, the lattice will immediately shrink.

There seems to be general agreement that all these effects are due to the changes in the long-range repulsive electrical forces between the filaments in muscle, as in all other colloidal systems.

The highly charged nature of the muscle filaments accounts for the fact that the filaments stand apart from each other in muscle and are stable in their lattice positions. As it happens, it also accounts for the constant-volume behaviour which we have already mentioned (Section 6.1.1). A fairly simple physical–chemical argument shows that, assuming that the total electrical charge on the filaments does not change, the volume occupied by one of the filaments should also not change and thus, as the overlap between the thin and thick filaments increases, the filaments must stand further apart. Details of this argument are beyond the scope of this Unit, but you can find them as an appendix to Matsubara and Elliott (1972). The argument can be extended to account for certain circumstances where the constant-volume condition no longer applies.

The colloidal view of muscle is important because it implies that there are regions of very high electrical potential gradient close to the highly charged filaments, that is, regions in which the electrical potential is changing swiftly, going from the surface of the filament into the inter-filament space. Close to the filament surfaces there are regions of high negative potential, and an incoming calcium ion will remain close to the filament surfaces because a calcium ion carries two positive charges in solution and will be strongly constrained to move in the most negative regions. On the other hand, the ATP ion which is multivalent and negative will tend to stay in the most positive region, which will be at the potential minimum that must exist somewhere between the filaments (Fig. 16(a)). It is probably for this reason that the myosin ATPase lies at the end of a projection which extends 5–10 nm into the mid-regions. When the calcium ion binds to the thin filament (Section 6.4.4) it will cause a local decrease in the negative potential of the thin filament surface or even, possibly, make the surface of the thin filament positive locally. This will cause a shift in the potential minimum towards the thin filament, or even a complete abolition of the minimum (Fig. 16(b)).

Figure 16(a) The inter-filament space between a negatively charged, thick-filament and a negatively charged thin filament, showing the potential gradient, and the complex of head.ADP. P_i held at the potential minimum by its net negative charge in the resting condition.

(*b*) When the charge on the thin filament is made locally more positive by the binding of Ca^{++} the head complex will move towards the thin filament.

N.B. These figures are highly diagrammatic, and the underlying idea is speculative.

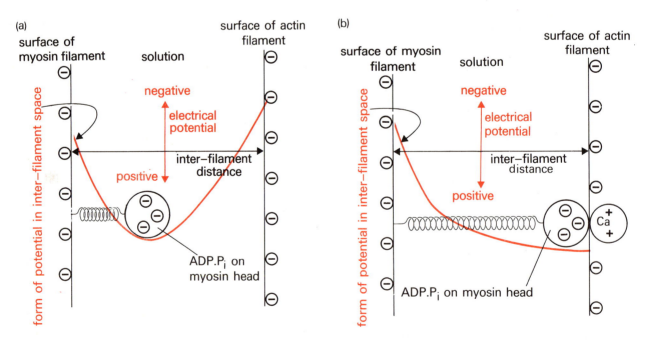

Under either circumstance the negatively-charged projections carrying the ATP will be moved towards the surface of the actin filament. In these ways, changes in the potential gradients may be controlling the contractile process.

The last two paragraphs must be regarded as speculative, but the general picture of the balance of the long-range forces seems to be accepted, although the details of the 'interactions' may need to be modified in the light of future work, both experimental and theoretical.

If you wish to follow the implications of this work further, you could start with Elliott (1973), which is a review-type article containing references to the original work.

6.4 The proteins of the contractile system

There are four main structural proteins in the contractile system of vertebrate muscles: myosin, actin, tropomyosin, and troponin.

In invertebrate muscle a further structural protein, paramyosin, occurs, but troponin is not always present. Myosin, actin and tropomyosin occur in all muscles so far examined, though probably these proteins are not precisely the same from all sources.

In this Section we shall collect a few data on the proteins, and on filaments constructed from them, which are relevant to our structural and functional interests. Even though we have given more detail than is required by our Objectives, we have had to omit much more again. In particular, we have not talked about the sub-units of myosin (light and heavy chains) or about the C protein, which is another protein of the thick filament. If you would like to find out more about these you can consult Offer (1973).

Notice that we give more detail than you need remember; re-read Objectives 2(c), 2(g) and 2(h) and Table A, before you proceed.

Table 1   The contractile proteins and their physical parameters

	Molecular weight	Molecular length* and diameter nm^{-1}**		Percentage of α-helix
Myosin	460 000	160	non-uniform	60
Light meromyosin	120 000	70	2	90
Heavy meromyosin	350 000	80	non-uniform	50
Tropomyosin	65 000	40	2	90
Paramyosin	200 000–220 000	140	2	90
Actin (globular form)	41 700	5	(globular)	probably about 20

* These lengths depend on the choice of model and, in the case of the meromyosins, on the conditions of digestion.

** Diameters are deduced from viscosity measurements, and are only very approximate.

Table 1 gives structural data on the four proteins and also on two fragments obtained by enzyme digestion of myosin; these are called light and heavy meromyosin. Myosin (and both its fragments), tropomyosin and paramyosin are all alpha proteins[10]: tropomyosin, paramyosin and light meromyosin are more than 90 per cent helical; heavy meromyosin, and myosin itself, have a region which is non-helical, consisting of two globular units (see below). Actin is not an alpha protein*, and does not give the X-ray diagram typical of such proteins.

alpha protein

Fibrous alpha proteins may consist of coiled coils[10] of two or three polypeptide chains (i.e. on the protofibril[10]). It seems probable that the muscle alpha proteins are all two-chain structures.

6.4.1 The myosin molecule and the products of proteolysis

The molecular weight of myosin has been measured by many investigators using several techniques, and several different values have been reported over the

* A point of definition arises here. An alpha protein, as originally defined by Astbury, is one which gives the characteristic alpha wide-angle X-ray pattern (see Appendix 1). Fibrous proteins which give this pattern have large amounts of alpha helix, arranged parallel to the axis of the fibres. But as the atomic structures of more and more proteins are solved, it has become clear that many proteins have some portions of alpha helix in their structure. The definition is thus a question of degree, and of arrangement.

years. Currently values around 4.6×10^5 seem accepted for myosin from rabbit skeletal muscle. Myosin can be split by brief tryptic or chymotryptic digestion into two major fragments, light and heavy meromyosin, together with some small peptides. The molecular weights of the two main products, given in Table 1, make up 80–90 per cent of the weight of the intact molecule, and there is general agreement that the two fragments are linearly arranged in the intact molecule, and are joined by a short region which is susceptible to the enzyme. Myosin can also be split by papain digestion. Papain intitially cleaves the myosin at the junction between the rod and the two heads. The rod and the individual heads can then be prepared in a relatively undegraded state. The individual heads are called 'heavy meromyosin subfragment 1', though we shall not generally use this term in Unit 6.

light meromyosin

heavy meromyosin

myosin head

Oriented preparations of light meromyosin give a good alpha high-angle X-ray diagram, and also low-angle reflections which suggest an axial periodicity distance of 42.8 nm. In intact muscle the myosin filaments, from the low-angle X-ray diagrams, also have a periodicity of 42.9 nm (as for light meromyosin) associated with cross-bridges or projections arranged on a helix round the filament axis, and also with the periodicity of the shafts of the thick filaments.

Structurally, the myosin molecule, seen in the electron microscope, is about 140–160 nm long, and consists of a thin rod about 2 nm in diameter with a thickened region at one end. Light meromyosin is rod-like, with no thickened region. The most recent study using a more advanced shadowing technique, has shown (see Fig. 1 on p. 7) that the thickened region is in fact double, consisting of two globular parts. These two parts are the two heads and bear the site of the myosin ATPase. Heavy meromyosin also contains a short portion of the rod-like rail, next to the heads (heavy meromyosin subfragment 2) and light meromyosin is the bulk of the tail of the myosin molecule. A schematic representation of the myosin molecule is shown in Figure 17, which gives more detail than Figure 1.

myosin rod (or tail)

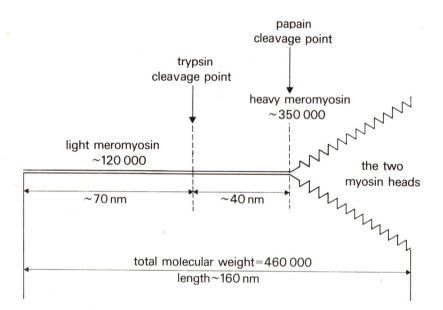

papain
cleavage point

trypsin
cleavage point

heavy meromyosin
~350 000

light meromyosin
~120 000

the two
myosin heads

~70 nm   ~40 nm

total molecular weight=460 000
length~160 nm

Figure 17 A more detailed representation of the myosin molecule, giving the probable lengths and molecular weights of the various fragments. The two-chain tail is shown separated, but is in fact twisted into a two-strand rope. We have also shown the points at which the molecule is cleaved by the enzymes trypsin and papain. The myosin heads have been shown diagrammatically as zig-zags. In reality the protein chain in each head forms a compact globular unit, probably twisting and turning back on itself many times. The details are not known.

As we saw earlier in Section 6.1.1, H. E. Huxley showed that when myosin molecules (from rabbit skeletal muscle) aggregate into myosin filaments at low ionic strengths, the first few molecules join in opposite directions with their tails partly overlapping (Fig. 1) and successive molecules then join on at the ends to give a bipolar filament one or two nanometers long with projections (the heads) along most of the length, but with a short region in the centre (0.15–0.2 nm long) without projections. These filaments have marked similarities to the thick filaments of striated muscle.

The effect of pH and ionic concentration on the genesis of these synthetic myosin filaments has been studied, and shows that the longest filaments occur at pH 6.5 and ionic strength 0.1M. (These are probably close to the *in vivo* conditions.)

6.4.2 Actin and actin filaments

The actin filament is a common feature of all the muscle contractile systems so far examined.

Actin filaments have been characterized structurally by moderate-angle X-ray diffraction and by electron microscopy. The X-ray diagram given by a muscle rich in actin filaments is shown in Figure 18, and the reflections from the actin

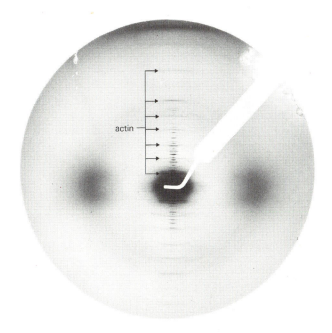

Figure 18 The X-ray diffraction diagram from a muscle rich in thin actin-containing filaments (an obliquely-striated muscle from an oyster, see Section 6.5.2). The diffuse actin reflections are marked. For further details see Appendix 1.

filaments are marked. We cannot, here, enter into the interpretation of such X-ray diagrams, but will note in passing that within the limit of experimental accuracy of measurement there has been uncertainty in the assignment of a long repeating distance in the filament direction. This distance ('the actin spacing') is the pitch of the helix, or the distance between cross-over points in Figure 3(b) (p. 8). It has been quoted as 41.5 nm and 36.0 nm, and the differences in observation which give these two apparently very different figures are in themselves rather subtle; in living resting muscle the latter figure is now generally accepted. Notice that this distance does not correspond to the repeat distance in the myosin filaments (42.9 nm). Each of the beads shown in Figure 3(a) and (b) is one actin molecule, an actin monomer.

actin molecule

The alternative structural approach, using the electron microscope, has given much additional information about actin. Hanson and Lowy have obtained detailed electron micrographs of actin filaments isolated from many types of muscle. As we have seen, these micrographs (see Fig. 3 on p. 8) suggest that alternate monomers are connected together into two helices, so that the structure is two-stranded and the monomers are very roughly spherical (diameter about 5.5 nm).

At the present level of interpretation, information about the shape and connection of the actin monomers is primarily available from the electron microscope studies, the X-ray studies not having developed to the point at which electron-density maps have been obtained. There are pointers that such a map for actin would be interesting. In the actin X-ray pattern, there is a striking intensity difference between two particular reflections which are close to each other. In other systems the origin of such differences appears to be the elongated and tilted nature of the scattering unit (that is, the unit is a diagonally placed egg, not a sphere). If the analogy is followed it seems likely that the scattering unit in actin may turn out to be elongated and tilted, perhaps across the two helices on Hanson and Lowy's model, rather than to be approximately spherical as suggested in Figure 3(b) (p. 8).

This shape is difficult to see with certainty in the electron microscope picture (Fig. 3(a)) although image-resynthesizing techniques are being used with success in H. E. Huxley's laboratory. In these techniques a computer is used to generate the most probable image from electron micrographs.

6.4.3 Tropomyosin

The actin structure deduced from X-ray and electron microscope studies fails to account for two phenomena associated with the thin filaments. These phenomena are: (a) a fine period of about 38.5 nm which is seen in the electron microscope in regions of the sarcomere where the actin filaments are located in striated muscle of vertebrates and (b) some of the X-ray reflections (particularly one at 38.5 nm) obtained from various muscles.

It has been suggested that these might both be manifestations of a combination within the thin filaments of actin and tropomyosin and the troponin regulatory system (see the next Section, 6.4.4). It had long been known that some methods for the extraction of actin produce a preparation in which there is 10–25 per cent of tropomyosin as a contaminant. Recently it has been shown that some preparations of tropomyosin also contained a further protein, troponin. It seems likely that all these proteins are closely associated *in situ*, as will become clear in the following Section.

6.4.4 Troponin and calcium activation

It has been known for some years that the injection of a minute quantity of calcium ions (10^{-6} M) into a single muscle fibre will produce a local contraction (see Fig. 19). At the same time, the (contractile) interaction of pure actin and myosin and ATP (MgATP) does not require the presence of the calcium ion.

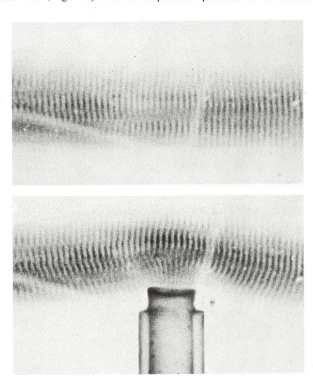

Figure 19 A local contraction in a skinned muscle fibre, produced by microinjection of Ca^{++} ions. ($\times$1 000)

Recently the answer to this paradox has become clearer, largely through the work of a Japanese scientist, Setsuro Ebashi, who has written a short one-page review of this work (Ebashi, 1972) giving references to the original papers.

The clue to the paradox is the occurrence, in the thin filaments in living muscle, of a regulatory system comprising the tropomyosin–troponin complex. Troponin is a protein which is arranged on the thin filaments at intervals of about 38.5 nm. It seems to be held in place by tropomyosin which lies in the grooves of the actin double helix (see Fig. 20). The calcium, released by the sarcoplasmic reticulum, reacts with the troponin. This seems to cause a small shift of position of the tropomyosin (a shift which can be detected by X-ray diffraction studies on contracting muscle). This shift is thought to uncover the site of interaction with the myosin-ATP system on the thin filament, so that the contractile process is initiated. Relaxation involves the sequestration of Ca^{++} back into the sarcoplasmic reticulum, the shifting back of tropomyosin and the interruption of the actin–myosin interaction.

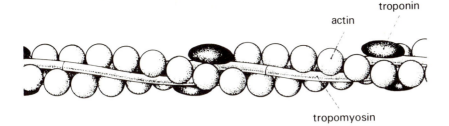

Figure 20 The tropomyosin–troponin complex lying in the grooves of the actin double helix; Ebashi's conception of the thin filament.

10 nm

In systems of pure actin and myosin, the tropomyosin–troponin regulatory system is not present, and the contractile process is not sensitive to the level of calcium.

As Ebashi says in his article, 'the subtle structural change that takes place in the thin filament—the troponin-tropomyosin-active complex—under the influence of calcium ions has become one of the vital problems in muscle science; the answer will undoubtedly provide a crucial clue about the interaction of myosin and actin'.

6.4.5 Paramyosin

It is necessary to begin with the history of the term 'paramyosin'. This word was first used in the structural sense to denote a particular periodicity and pattern seen in the electron microscope, and by X-ray diffraction (Fig. 21), in filaments from certain muscles. Later a new protein was isolated from these same muscles by Kenneth Bailey and was called 'water-insoluble tropomyosin' by him, to distinguish it from the tropomyosin we have already mentioned (Section 6.4.3), which Bailey also discovered and which he called 'water-soluble tropomyosin'. Later workers found that there were more differences between these two proteins than Bailey first supposed, and concluded that they were not homologous. These later workers also found the terminology clumsy, and started to use the name 'paramyosin' instead of 'water-insoluble tropomyosin'. This is a second and distinct use of the 'paramyosin', this time in biochemical terms. It is a pity that the name is now used in two distinct ways, because this can be confusing. It seems too late, however, to introduce a more logical terminology. We shall use paramyosin, unmodified, for the protein, and shall talk about the 'paramyosin structure' or 'paramyosin filaments' when we need the original definition. Notice though that the paramyosin structure, and paramyosin filaments, are not completely composed of the protein paramyosin, they contain myosin also. The probable structure of a paramyosin filament is a core of paramyosin molecules, packed to give rigidity and size, and a surface layer of myosin molecules, whose heads form the projections reaching out to the thin filaments.

Paramyosin has been found in animals of many invertebrate phyla, including the arthropods. It does not apparently occur in vertebrates. It has been reported as a major constituent, though, of the notochord of the cephalochordate *Amphioxus*.

The ratio of the amount of paramyosin to the amount of myosin varies greatly from muscle type to muscle type, from 1 : 10 at one extreme to 10 : 1 at the other. In insect flight muscle, or clam striated muscle (both fast muscles) the ratio is low, and in the smooth paramyosin adductor muscles (Sections 6.5.1 and 6.5.3) the ratio is high. On the basis of their physiology muscles with a large amount of paramyosin are often 'catch' muscles, which can remain contracted for long periods of time with little consumption of ATP (Section 6.5.3).

Unlike myosin, paramyosin has no ATPase activity. It is located in the core of the thick filaments; as we shall see the thick filaments are generally speaking longer and thicker in invertebrate than in vertebrate muscle.

Paramyosin is very like the rod portion of myosin (light meromyosin plus the helical part of heavy meromyosin). This similarity (see Table 1 on p. 19) is in

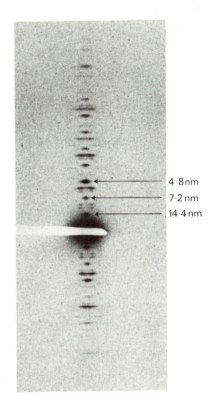

4·8 nm
7·2 nm
14·4 nm

Figure 21 The X-ray diffraction diagram given by the paramyosin filaments in oyster (smooth) muscle (see Section 6.5.3). The meridional reflection at 14.4 nm, 7.2 nm and 4.8 nm are marked.

size and shape, in amino-acid composition, in the high proportion of α-helix and in the 72.0 nm fundamental repeat period*.

It seems possible that paramyosin is an evolutionary development from the myosin molecule, specialized as a core protein for thicker and longer filaments, which might have some function in slow muscles. The head portion of myosin would make it difficult to pack in the core of a filament, but a protein related to the rod portion would not have this disadvantage.

Which came first, though? Here we can only speculate. Myosin in its headed form is present in very simple cells (*Amoebae*) so it is possible that myosin is the original protein and paramyosin is a later development in evolution.

6.4.6   The Lymm–Taylor scheme for actin–myosin interaction

As the contractile mechanism depends fundamentally on the hydrolysis of ATP by myosin, we would like to know the time course of this hydrolysis. Lymn–Taylor scheme Experiments have been performed where the enzyme and its substrate are mixed rapidly, and the time course of the production of inorganic phosphate is followed, using radioactive tracer techniques. It turns out that on mixing there is a rapid production of inorganic phosphate, and then a very small constant rate of further production. It appears moreover that this early burst of inorganic phosphate is not *released* by the enzyme but remains attached to it, apparently bound by secondary forces. These experiments, in the main, are due to Lymm and Taylor (see Offer, 1973, for further details). Lymm and Taylor explain their results by the following kinetic scheme.

$$M + ATP \underset{\longleftarrow}{\overset{fast}{\longrightarrow}} M.ATP \underset{\longleftarrow}{\overset{fast}{\longrightarrow}} M.ADP.P_i \underset{\longleftarrow}{\overset{slow}{\longrightarrow}} M + ADP + P_i$$

The enzyme (represented by M for myosin) combines rapidly with the substrate, which is hydrolysed to ADP and P_i (inorganic phosphate). These products are, however, bound to the enzyme and are released only slowly. In the presence of actin, the splitting of ATP in the second, slower, phase is increased about twenty-fold, though this is still about ten times less than the maximum rate of ATP splitting in striated muscle. It seems very probable that the precise arrangement of the muscle filaments is an important factor in achieving maximum ATP splitting. The maximum rate in striated muscle is about 35 molecules of ATP per myosin head per second.

Concluding note to Sections 6.4.1–6.4.6

It is difficult to summarize these Sections because, as we said earlier, they were designed to be a convenient collection of information about muscle proteins, and some of the filaments made from them. You are not expected to remember, much less to reproduce, the detail.

The amount of detail which you should keep in mind is that suggested by SAQs 9–17. These are part of a series of self-assessment questions (1–20) which try to draw out the main points of the part of Unit 6 which you have now read and to indicate the level of comprehension which is expected of you.

---

* This last similarity is subtle. The repeat period in paramyosin filaments is 72.0 nm, but there is a strong sub-period of 72.0/5, or 14.4 nm. The repeat period in myosin filaments is 42.9 nm, but there is a strong sub-period of 42.9/3, or 14.3 nm. It is this near-identity of sub-period which led Elliott and Worthington to suggest that the two types of filament might be made from very similar building blocks.

Self-assessment questions for Sections 6.1–6.4.6

SAQ 1 (*Objective 2(c)*) (a) The thick filaments in a striated muscle are made primarily of (select from (i)–(iv)):

(b) The thin filaments in striated muscle are made primarily of (select from (i)–(iv)):
   (i) myosin
   (ii) actin
   (iii) neither (i) nor (ii)
   (iv) both (i) and (ii)?

(c) Which set of filaments bears the cross-bridges:
   (v) the thick or
   (vi) the thin?

(d) Which set of filaments has a polarized structure (select from (vii)–(x)):
   (vii) the thick
   (viii) the thin
   (ix) both sets
   (x) neither set?

(e) During length changes of the whole muscle, which of the following statements is correct (select from (x)–(xiii))?
   (x) The A- and I-bands do not change length.
   (xi) The A-band changes length, and I-band does not.
   (xii) The I-band changes length, the A-band does not.
   (xiii) Both A- and I-bands change length.

SAQ 2 (*Objective 2(c)*) Draw a diagram showing the basic structure of:

(a) an actin filament;

(b) a myosin filament.

SAQ 3 (*Objective 2(d)*) Twitch tension is less than tetanic tension because (select from (i)–(iii)):
   (i) during a twitch the contractile component does not have time to develop full tension
   (ii) during a twich some of the work done by the contractile component goes to overcome series elasticity
   (iii) during a twitch not all of the muscle fibres are activated?

SAQ 4 (*Objective 2(e)*) Draw a diagram to show how the tension is related to the degree of overlap of the thick and thin filaments. Write a few sentences explaining the different phases of the diagram you have drawn, in terms of the actin–myosin interaction.

SAQ 5 (*Objective 2(a) and (b)*) Does a muscle contract most quickly (select from (i)–(iii)):
   (i) under full isometric load
   (ii) under zero load
   (iii) when the load is at some optimum between (a) and (b)?

SAQ 6 (*Objective 2(f)*) The prime function of the transverse tubules is (select from (i)–(iv)):
   (i) to hold the myofibrils in place
   (ii) to provide an energy story for contraction
   (iii) to transmit surface depolarization into the muscle fibre
   (iv) to house the enzymes which make the contractile proteins?

SAQ 7 (*Objective 2(f)*) The immediate trigger of contraction in a muscle fibre is (select from (i)–(v)):
   (i) the synthesis of ATP in the fibre
   (ii) the breakdown of glycogen in the mitochondrion
   (iii) the release of the calcium ion into the myofibril
   (iv) the mechanical deformation of the myosin molecule
   (v) the formation of actin filaments from globular molecules?

SAQ 8 Does a muscle fibre produce a force when it relaxes, opposite to the force produced when it contracts?

SAQ 9 (*Objective 2(g)*) Which of the following proteins (or fragments) are concerned primarily with the control of the contractile event (select from (i)–(vii))?

   (i)   heavy meromyosin
  (ii) light meromyosin
 (iii) actin
 (iv) tropomyosin
  (v) paramyosin
 (vi) troponin
(vii) myosin.

SAQ 10 (*Objective 2(g) and (h)*) Which of the proteins, from the list given in SAQ 9, excluding item (vi), has ATPase activity?

SAQ 11 (*Objective 2(g)*) Which of the proteins, from the list given in SAQ 9, is not an alpha protein (i.e. does not give an alpha X-ray diagram)? Of those that are, which have the highest amount of alpha helix?

SAQ 12 (*Objective 2(g)*) Which of the proteins, from the list given in SAQ 9, forms filaments whose structure is the same in a wide variety of different smooth and striated muscles?

SAQ 13 (*Objective 2(g)*) Which of the proteins, from the list given in SAQ 9, is located in the thin filaments, in addition to actin?

SAQ 14 (*Objective 2(g)*) Which of the proteins, from the list given in SAQ 9, does not occur in vertebrate muscle, as far as we know?

SAQ 15 (*Objective 2(g)*) Which of the proteins, from the list given in SAQ 9, are essential for the production of a contractile force?

SAQ 16 (*Objective 2(g)*) Which of the proteins, from the list given in SAQ 9, has the highest molecular weight?

SAQ 17 (*Objective 2(a) and (b)*) Why is the constant-volume phenomena significant in considering models of contraction?

SAQ 18 (*Objective (2a) and (b)*) During muscle contraction, the part of the X-ray diffraction pattern that arises from the myosin projections (the 'cross-bridges') becomes much less intense. Why is this significant?

SAQ 19 (*Objective 2(a) and (b)*) Do the basic periodicities of the actin and myosin filaments (as measured by X-ray diffraction) change during contraction?

SAQ 20 (*Objective 2(a) and (b)*) How is the speed of contraction of a muscle related to the external load on the muscle?

6.5   Other types of muscle

In Sections 6.1–6.1.5 you met the most highly ordered muscle, vertebrate striated muscle, and you learned about the sliding-filament model. You were left with the fundamental question about the mode of action of the sliding-filament system—what makes it slide? Now there is not yet a generally agreed answer to this question but it seems inconceivable that the basic contractile event will not be the same in all types of muscle, not just in striated muscle. It seems likely that it will be the same or at least similar in other contractile systems, in cilia, flagella and so on, unless evolution has given rise to several quite different systems.

We now go on to survey the manifold forms of contractile system, in an effort to find the underlying similarities, because this may give clues to the basic contractile event. Less is known of these other contractile systems than is known of striated muscle but there is nevertheless much to illuminate our path. After this survey we shall return to the basic questions to see whether our appreciation has sharpened to the point of giving answers.

Any attempt to classify all muscles will probably turn out to be premature. The muscles which will be considered in Sections 6.5–6.5.4 fall into three convenient groups; obliquely-striated muscles, paramyosin smooth muscles and smooth muscles of vertebrates.

obliquely-striated muscle

Obliquely-striated muscles are widespread; they occur particularly in the body muscles of echinoderms, annelids and molluscs (cephalopod, gastropod and lamellibranch) but not, apparently, in arthropods or vertebrates.

paramyosin smooth muscle

The fibres of these muscles show a 'double-oblique striation' in the light microscope; this results from the appearance of both aspects of some helical structure within the fibre (Fig. 22). These muscles are a special case of striated muscles, but all of them so far examined have thick filaments of the paramyosin type (Section 6.4.5) rather than of the myosin type found in the striated muscles of vertebrates and arthropods.

vertebrate smooth muscle

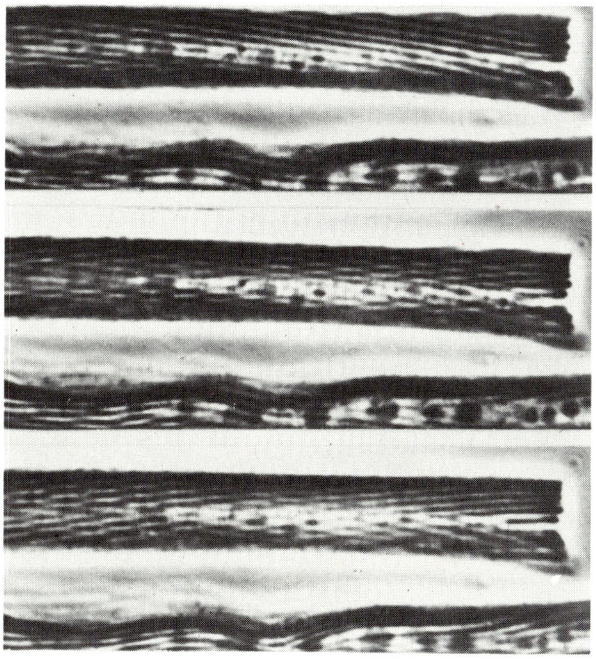

Figure 22 Phase-contrast micrograph of obliquely-striated muscle—squid mantle muscle. The end of a broken fibre at three focal levels, showing how the oblique striations are aspects of a basic helical structure. (×1 500)

Paramyosin smooth muscles occur particularly in molluscs and annelids. In these smooth muscles the fibres contain longitudinal filaments of large size (the paramyosin filaments) sometimes more than 0.1 μm in diameter. These filaments have the characteristic paramyosin structure, readily detected by observations in the electron miscroscope or from the X-ray diffraction pattern (Appendix 1) of the intact muscle (see Fig. 21 on p. 23).

The smooth muscles of vertebrates occur as certain involuntary muscles, for example, vascular muscles (the blood vessels) and visceral muscles (gut and uterus muscles). (The bladder is a vertebrate smooth muscle which *is* under voluntary control.)

There are many other muscles in organisms throughout the animal kingdom which are described as 'smooth' in a negative sense, because they lack transverse striations when examined in the light microscope. More work would be necessary before the classification could be more precise.

6.5.1 The adductor muscles of bivalve molluscs

The adductor muscle of a bivalve mollusc usually contains two parts, one used to hold the shell closed and one to close the shell swiftly. The muscle used to close the shell swiftly is sometimes a cross-striated muscle, for example in *Pecten*, but is more often an obliquely-striated muscle. The muscle used to hold the shell closed is usually a paramyosin smooth muscle. A good example of this differentiation is the oyster. The adductor muscle can be seen to consist of two parts, one opaque and white and one pink and translucent (Fig. 23). The opaque

Figure 23 An oyster, dissected to show the two halves of the adductor muscle. The obliquely-striated portion is darker, on the left in this picture.

adductor muscle is a muscle which under certain conditions of stimulation has a relaxation rate (called the tonic rate) which is one hundred times slower than the rate under different stimulation conditions (called the phasic rate). This muscle is a 'catch' muscle, that is, it is specialized to maintain contractions for long periods with infrequent stimulation and a low rate of consumption of ATP (i.e. low energy consumption). ('Catch' responses were first demonstrated in a similar muscle of another mollusc, *Anodonta*, by Pavlov.) Structurally the opaque adductor muscle is a paramyosin smooth muscle. The translucent adductor muscle of the same animal has also been studied mechanically and structurally. This muscle does not give a 'catch' response; it is the muscle which the animal uses to close the shell swiftly and is an obliquely-striated muscle. Both the parts of this muscle contain paramyosin filaments. The translucent part will be discussed in Section 6.5.2 and the opaque part in Section 6.5.3. In these molluscan muscles the calcium-regulatory role is not, it seems, controlled by a tropomyosin–troponin system on the thin filaments (Section 6.5.4). It is, apparently, concerned directly with the thick myosin filaments, and involves the myosin light chains which we mentioned in passing in the study comment at the beginning of Section 6.5.

catch muscle

6.5.2 Obliquely-striated muscles

These muscles are detected by observations in the light microscope where they are seen to have some structural elements arranged helically in the fibre (Fig. 22). The early microscopic observations were generally made with fixed and sectioned material and could not always be interpreted without ambiguity. More recently, observations on intact glycerinated fibres, using the phase-contrast microscope, have shown beyond doubt that the helical structure is not an artefact and, in particular, that it is not a distorted cross striation.

There seem to be several variations of the helical arrangement. In one type the fibre is ribbon-shaped so that opposite sides of the helix are closely apposed.

An example of this variation is earthworm body-wall muscle. Another variation has a central core of material, containing many mitochondria, surrounded by the contractile material. Examples are cuttlefish and squid mantle muscles.

As will be seen in detail later (Fig. 24) the helical nature of these muscles is a consequence of an arrangement of the thick and thin filaments. In obliquely-striated muscles the A-band filaments are arranged helically around the fibre, and this gives rise to the double-oblique appearance.

In the nineteenth century it was observed that when obliquely-striated muscles contract there is a change in the inclination, with respect to the fibre axis, of the helical bands. In squid mantle muscle the change is from 5–10° to as much as 60°. In a striated muscle with helically-wound A-bands any shortening of the sarcomere will automatically produce a change in the angle made by the bands with the fibre axis, supposing that the muscle works at constant volume, as do other muscles.

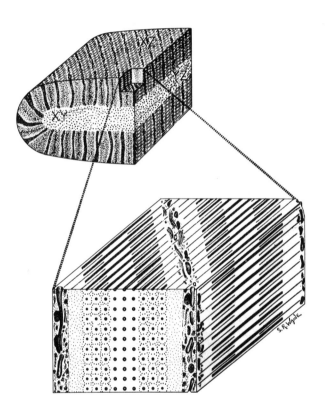

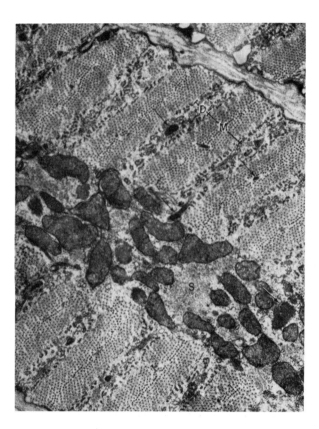

It is not known what general physiological significance can be attached to the helically-wound A-bands in these muscles. It has been suggested that the 'shearing' of filaments possible in obliquely-striated muscle could make it possible for the muscle to change lengths by larger amounts. In the case of a sufficiently long muscle fibre with regular A-bands and I-bands, however, for a given change in sarcomere length the change in muscle length does not depend appreciably on whether or not the A-bands are helically wound.

In the past there has been some argument over the direction of the filaments in certain of these obliquely-striated muscles (particularly in the squid muscles), but there now seems general agreement that, despite the helical appearance, the filaments are in all cases parallel to the fibre axis.

Figure 24 is a generalized diagrammatic representation of the filament arrangement in an obliquely-striated muscle, as revealed in the electron microscope. (Compare this diagram with Fig. 25, a cross-section of muscle from the worm *Ascaris*.) This arrangement, with minor modifications, applies to the outer region of the cell in both the variations which we mentioned above. The helical appearance seen in the light microscope is a consequence of the arrangement of the two sets of filaments, thick and thin, in a manner analogous to cross-striated muscle. The thin filaments seem to be connected to dense bodies which appear to be analogous to the Z-line structure in cross-striated muscle. Figure 26, a longitudinal section of oyster muscle, shows these dense bodies.

Figure 24 (left) General diagram of the filament attangement in an obliquely-striated muscle. The central core is surrounded by helically-wound A-bands. The insert shows the filament arrangement in more detail.

Figure 25 (right) Electron micrograph of a cross-section of a muscle fibre from the worm *Ascaris*, an obliquely-striated muscle, for comparison with Figure 24. Notice the core substance (marked S), containing many mitochondria, the A-band regions (marked M) and the I-band and dense-body regions (marked D). (× 15 000)

Under high magnification the two sets of filaments may be distinguished clearly where they overlap; the thick filaments show the 14.4 nm banding, associated with the paramyosin structure (this is the paramyosin sub-period, see the footnote on p. 24) and taper to a point at either end. There are regions where the thin filaments stand alone, regions where the two sets of filaments overlap and, in stretched muscle, regions where the thick filaments are not overlapped by thin

Figure 26 Low magnification electron micrograph of a longitudinal section of the obliquely-striated muscle from the oyster, showing the two sets of filaments, the A-band regions and the I-band regions containing the dense bodies. ($\times$ 16 000)

Figure 27 Electron micrograph of longitudinal sections of the obliquely-striated earthworm adductor muscle, seen at higher magnification ($\times$ 26 000)

filaments (Fig 27). These regions are analogous to the I-band, overlap zone, and H-zone of cross-striated muscle. It can be shown that the amount of overlap increases as the muscle shortens, with the eventual disappearance of the I-band and H-zone analogues. Thus the muscle shortens by a sliding-filament mechanism. The A-band (H-zone and overlap zone) is about 5 μm long in the oyster translucent muscle, at the periphery of the fibre. The two sets of filaments appear to be connected by cross-bridges in the overlap region.

30

The filaments of the oyster translucent adductor muscle (among other types) have also been examined by negative-staining methods (see Unit 1) after isolation from homogenized preparations (Fig. 28). The thin filaments have the characteristic actin structure (Section 6.4.2) and the thick paramyosin filaments taper to a point at either end, bearing projections right up to those points. In thinking about the significance of this you should remember the point that was made in Section 6.4.5. Paramyosin filaments are structurally related to myosin filaments and, in particular, they have a surface coating of myosin molecules with heads (the projections in Fig. 28) reaching out towards the actin filaments.

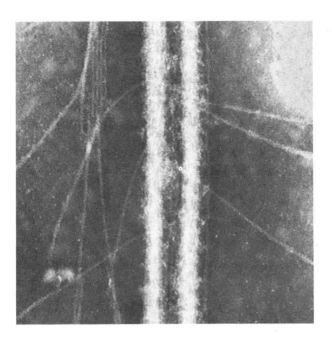

Figure 28 Electron micrograph of negatively stained preparations of the filaments from oyster translucent adductor muscle, at high magnification, showing the thin actin filaments with their characteristic structure (compare Fig. 3(a)) and the projections on the thick filaments reaching out to the thin. (× 160 000)

Summary of Section 6.5.2

In summary, obliquely-striated muscles are a special case of cross-striated muscle where the A-bands are arranged in a large-scale helix. The physiological reasons for the oblique striation, and for the occurrence in these muscles of thick filaments of the paramyosin (rather than the myosin) structure is not known.

6.5.3 Paramyosin smooth muscle

The opaque muscles of the clam and the oyster are typical examples of paramyosin muscles that are *not* obliquely-striated, as they show no helix when examined in the light microscope (Fig. 29). These smooth muscles also have two sets of filaments, thick and thin. The thin filaments contain actin, giving the usual actin X-ray diagram and showing the characteristic structure in negatively-stained preparations of isolated filaments. Transverse sections show that the thick paramyosin filaments are up to 0.1 μm in diameter in these muscles and are roughly cylindrical in cross-section, though often rather irregularly shaped. In negatively-stained preparations of isolated filaments the intact paramyosin filaments have been shown to be 10–30 μm long, to be tapered at both ends and to bear projections as in other muscles. It is not known whether the thick filaments in a given muscle cell are all equal in length, because the filaments observed in one preparation come from many cells.

Measurements have been made of the thick filament diameters and of the number of thick filaments in unit cross-sectional area in pairs of muscles, one of which was fixed at normal length, and one stretched threefold and then fixed. It was found that the distribution of diameters of the filaments did not differ significantly and that there were about the same number of thick filaments per unit cross-section at the long and short lengths. If the muscle cell works approximately at constant volume, as is likely (Section 6.1.1), the total cross-sectional area of the cell is three times as much at the short length, so the total number of thick filaments per cross-sectioned cell has increased threefold. As in vertebrate striated muscle, X-ray diffraction experiments have not detected any change in

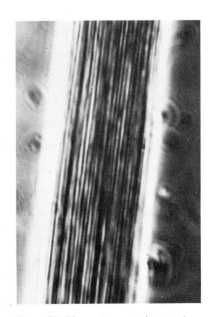

Figure 29 Phase-contrast micrograph of paramyosin smooth muscle—a single fibre from the opaque adductor muscle of the oyster. (× 2 500)

spacing of the molecules in the thick and thin filaments as the muscle length changes. Taken together these show that the thick filaments are discontinuous, and slide past each other as the muscle changes length. In the electron microscope, it can be seen that there are A-band regions, where thick and thin filaments occur together, and I-band regions, where only thin filaments and dense bodies (see the previous Section) occur. We may deduce that these muscles are structurally similar to the obliquely-striated type, with 'A-bands' that are much longer and very much less well-ordered.

There have been few published electron microscope cross-sections which show very clearly how the thin filaments are arranged in these muscles. Occasionally, rings of thin filaments have been observed around the thick filaments. These rings contain a large and variable number of thin filaments, up to about 12. Other electron microscope studies have shown other regions where the thin filaments, in the absence of thick filaments, are arranged in rough hexagonal array.

In longitudinal sections of these muscles the thin filaments lie closely alongside the thick, with the appearance of cross-bridges (Fig. 30). This close association of thick and thin filaments is apparent also from work on negatively stained homogenates of paramyosin smooth muscles.

Figure 30 Electron micrograph of section of oyster opaque adductor muscle, fixed at the length in the shell when the shell is closed. Note the thin filaments (arrowed) and the variety of band patterns on the thick filaments. ($\times 55\,000$)

An account of paramyosin smooth muscles would not be complete if it did not deal with the 'catch' function. This slowly relaxing (tonic) contraction of these muscles, with low energy consumption, first studied by Pavlov, has already been described (Section 6.5.1). Functionally, the response serves to keep the shells of the animal closed against sustained force exerted by a predator, etc. There are two distinct hypotheses to account for the 'catch' and the hypotheses differ mainly in the role ascribed to paramyosin.

The first hypothesis supposes that paramyosin alone is responsible for the maintenance of tonic tension. Tension developed by an actomyosin system, is maintained by a stage of rigidity in a parallel system of paramyosin (Fig. 31).

Now it is true that the muscles which can best maintain tonic contraction are those that have large thick filaments and have a high paramyosin to myosin ratio (Section 6.5.4). The correlation between the amount of paramyosin and the size of the thick filaments can therefore be interpreted as a correlation between the amount of paramyosin and the ability to maintain the tonic state. If it could be shown that the thick filament was continuous from one end of the muscle to the other it would be tempting to think that the 'catch' might be the result of

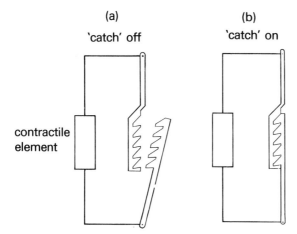

(a) 'catch' off

(b) 'catch' on

contractile element

Figure 31 The parallel ratchet hypothesis for 'catch' muscles. (*a*) The contractile system shortens the muscle and then (*b*) the ratchet closes, making the muscle strongly resistant to extension even if the contractile system is no longer active.

extra bonding (inside the thick filaments) between molecules which, in the relaxed state, could slide relative to one another. However, in 'catch' muscles the thick filaments are discontinuous, and slide past each other as the muscle shortens. In view of this it is necessary, in order to uphold the 'catch' hypothesis of an independent parallel system, to postulate rather implausible extra forces or bonds between paramyosin filaments to give structural continuity to paramyosin throughout the cell.

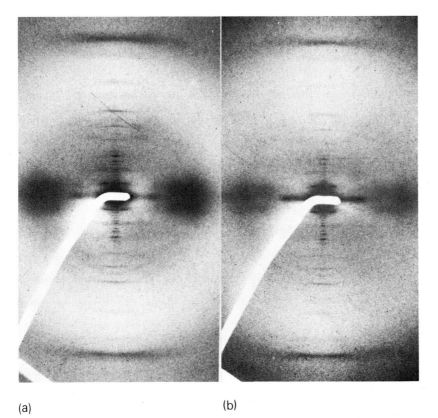

(a)

(b)

Figure 32 Comparison of X-ray pictures of a 'catch' muscle resting and during the long-lasting contraction.

(a) At rest.
(b) During tonic contraction.

Recently there have been comparisons between the low-angle X-ray patterns of such muscles when relaxed and when contracting tonically (Fig. 32). No major changes in spacing or intensity occur between the two patterns, and the conclusion is that there is no overall change in the arrangement of the molecules within either the actin or the paramyosin filaments. Thus there seems no major structural change associated with the 'catch' state.

The alternative hypothesis supposes that forces are set up between the actin filaments and the myosin molecules on the surface of the thick filaments, and that these forces operate a sliding-filament system. The tonic maintenance of

33

tension is a result of the very slow decay of these forces in the absence of a relaxing system. The activation of this relaxing system results in fast phasic relaxation of the muscles.

We cannot give, in the space available, more than this brief discussion of the two hypotheses. The evidence from measurements of the mechanical properties of living muscles favours the second hypothesis, while the evidence from glycerinated muscles has been interpreted to support the first, or 'independent catch' concept. The structural considerations are very important in the disposal of the controversy.

Summary of Section 6.5.3

Paramyosin smooth muscles are an extension of the obliquely striated type, but the A-bands are longer and less well ordered, so that they cannot be observed by light microscopy. Many paramyosin smooth muscles are also catch muscles, relaxing extremely slowly under certain conditions, with very low ATP consumption. There seems little or no evidence for a special structural mechanism for this slow relaxation.

6.5.4 Smooth muscles of vertebrates

Muscles in this group which have been studied in the electron microscope in thin sections include vertebrate vascular and visceral muscle. Certain generalizations apply to the results of all these studies: the muscle cells are long and spindle shaped, with a central nucleus, and are packed with filaments, which lie approximately parallel to the long axis of the cell. Some of these filaments are about the same diameter as the actin filaments in striated muscle (5–8 nm). The cells contain irregular dense bodies, elongated longitudinally, which lie among the filament mass (Fig. 33) and there are also mitochondria and a number of vesicles. For some time these muscles were thought to be unique, having only a single set of filaments—as far as could be seen in the electron microscope in intact, sectioned material. (Although in the soluble form both actin and myosin protein could be obtained from the muscles.)

Figure 33 Low magnification electron micrograph of a longitudinal section of visceral smooth muscle (taenia coli of the guinea pig), showing the central nucleus and the dense bodies lying in the filament mass. (× 7 000)

As it is possible to recognize certain X-ray reflections as arising from filaments of known type (Appendix 1), the diffraction method can also be used to diagnose filament types. The first such studies of these muscles (generally the taenia coli, an intestinal muscle of the guinea pig) showed a moderate-angle pattern typical of actin and low-angle reflections typical of collagen, but no reflections at low angles which might have come from myosin filaments.

In later observations of taenia coli muscle during peristaltic contraction it was again not possible to detect myosin reflections. Recently, in J. Lowy's laboratory

34

in Aarhus (Denmark) a clear 14.4 nm meridional reflection, diagnostic of myosin filaments, has been seen from stretched taenia coli muscle (Fig. 34). The intensity of this reflection changes on rhythmic (peristaltic) contraction of the muscle, confirming (by analogy with striated muscle) that the reflection does arise from the myosin component.

In an electron microscope study of thin sections of smooth chicken gizzard muscle made about 10 years ago, darker, thicker filaments were first seen by Jae Kwon Choi. These filaments, which will here be called 'dark filaments', could be distinguished from a second set of smaller filaments, which presumably were the thin filaments previously reported in smooth muscles of vertebrates.

Kate Shoenberg at Cambridge (England) studied the conditions under which these dark filaments were seen most clearly, in sections of chicken gizzard and guinea pig taenia coli muscle. She found that the dark filaments were usually present in fresh-fixed muscle (Fig. 35(a)) but were absent or less clearly visible when the cell membrane had been injured, and the injury followed by prolonged washing in solutions of low ionic strength (0.1 M) before fixation (Fig. 35(b)).

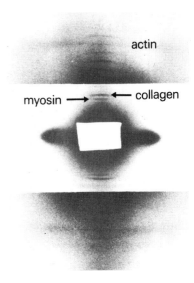

Figure 34 X-ray diffraction pattern from living taenia coli muscle, showing the 14.4 nm myosin reflection, the 13.0 nm collagen reflections and the 5.9 nm actin reflection.

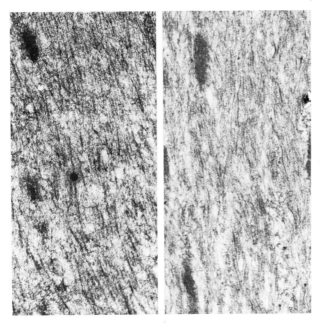

Figure 35(a) (left) Electron micrograph of section of fresh-fixed chicken gizzard muscle, showing clear dark filaments. (×50 000)

(b) (right) Electron micrograph of section of chicken gizzard muscle, glycerinated before fixation to destroy the cell membranes. The dark filaments are less distinct, and less numerous than in the fresh-fixed muscle. (×50 000)

Two possibilities were suggested; that the dark filaments consisted of myosin of high solubility and/or of actin filaments on which myosin had been deposited. In either case, supposing that a sliding-filaments system operates, the dark filaments are presumably the analogues of the myosin filaments in cross-striated muscles.

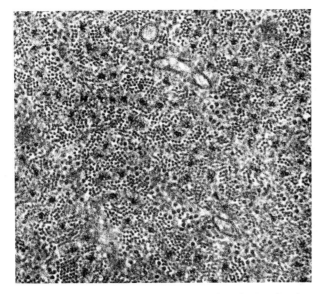

Figure 36 Thick and thin filaments seen in cross-section of taenia coli muscle fixed during contraction. (×74 000)

A group of workers in R. V. Rice's laboratory in Pittsburgh observed dark filaments in sectioned glycerinated chicken gizzard muscle fixed at low pH (pH below about 6.6). They later studied taenia coli muscle and chicken gizzard muscle which had not been glycerinated, and observed two sets of filaments in certain (but not all) cells fixed in contraction (Fig. 36). Relaxed muscle contained only thin filaments. They concluded that smooth muscle contracted by a sliding-filament mechanism, but that the thick filaments are in a highly labile condition. Later A. P. Somlyo in Philadelphia noticed a regular arrangement of thick, dark, filaments in vascular smooth muscle, both relaxed and contracted, and Rice's laboratory concurred with this observation. The dark filaments were spaced at intervals of about 70 nm (Fig. 37).

Figure 37 Cross-section of a muscle cell from the vein of a rabbit. The thick filaments are arranged with a popular spacing of around 70.0 nm. (× 70 000)

The variant results of the thin-sectioning techniques, and the sparse appearance of the dark filaments when they are seen (though this is not always the case, see Fig. 37) leaves room for speculation on the resting condition of the myosin component. The equatorial X-ray reflection at 12.0 nm seen with taenia coli shows beyond doubt that there must be areas of fairly regular packing of the actin filaments in these muscles in the living, physiologically resting condition. Until the electron microscope fixation and other techniques have advanced routinely to this point, it is probably still wise to treat the 'dark filaments' with some caution. The most convincing evidence for the existence of myosin filaments *in vivo* (resting muscle) is the X-ray meridional reflection seen by the Aarhus group.

Actomyosin, the complex of actin and myosin, can be extracted from vertebrate smooth muscle at lower ionic strengths than from striated muscle (i.e. it is more soluble). After extraction, as the result of treatments which possibly result in the loss of some solubilizing factor, this actomyosin can acquire behaviour indistinguishable in viscosity and ATP sensitivity from that of striated muscle actomyosin. Little difference is found between the actin components of the actomyosin from smooth and striated vertebrate muscles, and it is the myosin partner of the actomyosin which shows these lower solubility characteristics, and also shows large differences in amino-acid composition.

actomyosin

In physical–chemical respects, vertebrate smooth muscle actomyosin resembles skeletal muscle actomyosin, although ATPase activity is only one-tenth of that of skeletal muscle. It is possible, however, to prepare from uterus muscle a complex of actin and heavy meromyosin which has an ATPase activity comparable to that of the equivalent striated muscle preparation. Dorothy Needham, one of the most distinguished biochemists in this field, remarks that some unmasking of potentially active centres must have taken place, but that it is not yet known whether this can also take place under physiological conditions.

Another line of attack on the location of the contractile proteins in the vertebrate smooth muscle cell has been the study of homogenized preparations of the filaments, seen with negative-staining techniques in the electron microscope (see Unit 1). Typical actin filaments (Section 6.3.2) were observed in such homogenates, but it was not at first possible to find any other filaments. Later, actomyosin from vertebrate smooth muscle (rat uterus muscle) was disassociated with ATP to give myosin solutions. When these solutions were treated in the same way as similar solutions from vertebrate skeletal muscle, and examined in the electron microscope, short filaments were seen (Fig. 38) which had the characteristic structure of myosin filaments.

The Pittsburgh group obtained myosin filaments from chicken gizzard muscle homogenized in ATP-salt solution at pH 6.8. These filaments were about 13 nm in width, and 0.35 to 0.5 μm long. It is possible that these filaments were precipitated from solution, rather than originally present in the muscle in filamentous form. Shoenberg made a detailed study of the conditions under which the myosin filaments were formed *in vitro*, and concluded that Mg^{++}, Ca^{++} and ATP were all necessary for myosin filament formation (at pH 6.85). She suggested that the myosin is always present but only aggregates into filaments at the onset of contraction when Ca^{++} is released.

In the light of all these observations, the status of the myosin filaments in vertebrate smooth muscle seems still in doubt. The absence of a clear, ever present, myosin X-ray pattern is puzzling as the degree of order and orientation of the actin and myosin filaments should be similar, and a clear, well oriented actin pattern is observed at all times. It has been postulated that the myosin is present, at least in the resting muscle, as a soluble form which in fresh muscle aggregates or is deposited on the actin filaments during fixation. It would be most useful if these points could be clarified. If these muscles are a special case of sliding filaments, despite the apparent dissimilarities, detailed understanding of their contractile structure would very likely give useful clues to the nature of the fundamental contractile event. If, on the other hand, these muscles are not a special case of sliding filaments, having no resting analogue of the myosin filaments, attention will be focused on the actin filaments as a basic engine of contraction.

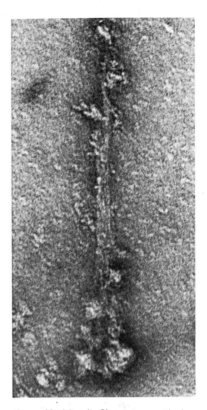

Figure 38 Myosin filament, negatively stained, obtained from homogenized taenia coli muscle by dissociation of the soluble actomyosin. ($\times 27\,000$)

Summary of Section 6.5.5

Vertebrate smooth muscle contains a mass of thin actin-containing filaments, but the existence of myosin-containing filaments is not yet established beyond doubt in the resting muscle. Probably these muscles are another case of a sliding-filament system with much less well-ordered A-bands, but it remains just possible that the myosin component is soluble in the resting muscle, and only forms filaments on activation.

6.6 Non-muscular contractile systems

There are myriads of contractile and motile processes in cell biology which do not involve specialized muscle cells. For example, there are cilia and flagella, the centriole and the spindle, which seems to move the chromosomes during cell division, and microtubules which appear to move materials within individual cells. Recently actin filaments—identified by their double helical structure (Section 6.4.2)—have been identified in several types of cells. These include an *Amoeba*, at one end of the scale of complexity and the glial cells of the human brain at the other end.

As a further aid to diagnosis, the filaments (thought to be actin) can be reacted with the globular myosin fraction (heavy meromyosin) and shown to form the characteristic 'arrow head' complexes (Section 6.4.2). These complexes are specific to actin filaments, and are not formed by any other filament found in tissues (collagen filaments, neurofilaments, etc.). Figure 39 shows these decorated actin filaments formed by filaments isolated from slime mould (when reacted with heavy meromyosin made from rabbits). Slime mould, which probably arose early in evolution, exhibits vigorous cytoplasmic streaming. Now cytoplasmic streaming (see Unit 1) is a widespread phenomenon in both plant and animal cells, and judging by its distribution in the phyla it probably developed early in evolution, almost certainly before muscle contraction. This would suggest that muscle contraction may have evolved from a primitive form of cytoplasmic streaming. Just how this illuminates the primitive contractile event is not clear at present, but it seems that this is a promising avenue for further study.

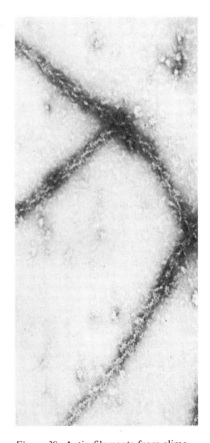

Figure 39 Actin filaments from slime mould, decorated with (rabbit) heavy meromyosin. ($\times 150\,000$)

6.6.1 Microtubules and cilia   (See also Unit 1, Section 1.4.8)

In a wide variety of situations in cells of many types there are long, hollow, cylindrical structures. These have been called 'micro-tubules'. They are usually about 23 nm in diameter, very long and straight, and are made of globular sub-units. Microtubules in neural tissues are illustrated in Figures 32 and 33 of Unit 1. In animal cells microtubules may occur through the cytoplasm, apparently at random, or they may be organized into higher level structures—the cilium, sperm tails and the centriole, etc. They also occur in higher plant cells, in the centriole and elsewhere. Other contractile organelles, such as the 'myoneme' and the 'kinetosome' identified in the light microscope, are also composed of microtubules, it appears.

microtubules

cilia

Of course, there is no *proof* that such structures, seen in the electron microscope, are strictly homologous from cell type to cell type. However, an equal degree of homology is readily accepted for other cell organelles (mitochondria, Golgi apparatus) and we can, as a working hypothesis, accept that all straight slender cylinders of this kind constitute a class of cell component.

Microtubules occur in very great numbers in protozoa; axostyles, for example, are sometimes a regular array of microtubules and protozoan flagella have the 9 + 2 arrangement of microtubules (see below). Microtubules are ubiquitous throughout the animal and plant kingdom, sometimes transient (sperm tails, centrioles) and sometimes permanent (cilia, nerve cell processes).

Microtubules, particularly in their transient manifestations, dissolve in response to low temperatures and high hydrostatic pressures. This dissolution is reversible. They also seem to be sensitive to pH changes; lower pH causes greater association which suggests some kind of electrical forces between the components and the tubules themselves. It has been suggested that a pool of macromolecules (tuberculin molecules, see Unit 1) exists in all cells, and that under appropriate physical–chemical conditions these molecules can associate to form micro-tubules, which then go on to form their various functions. What then are the functions?

The following list has been suggested:

1   To develop and maintain the form of asymmetric cells (nerves for example).

2   To provide limiting channels for the movement of the cytoplasm.

3   To provide the motive force for the streaming motion of the cytoplasm, which takes place relative to the more stationary microtubular arrays.

4   To provide a passive structure for the attachment of an actin-based contrac-tile system (in analogy with bone and muscle).

This is a list which raises many questions, and there are very few definite answers.

Going on to consider the structures in which the microtubules are arranged in some definite pattern, most is known about cilia, protozoan flagella and sperm tails, which can be considered together on structural grounds. These (which for convenience from now on we shall refer to as cilia) are long cylinders which taper to a blunt point and are membrane-bound projections from a cell. The diameter of the cylinder is about 0.2 μm, and the length can be from about 5 μm in some species to about 15 μm in others. Cilia were recognized in the early days of light microscopy, but nowadays are best defined morphologically by the 9 + 2 arrangement of microtubules which is seen in Figure 40. Each of the outer nine is a double microtubule. This arrangement obviously is very similar to the structure of the centriole (Unit 1, Section 1.4.7).

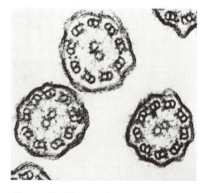

Figure 40 The 9 + 2 arrangement of microtubules in a cilium. Notice the arms on the outer microtubules. ($\times$ 65 000)

The 9 + 2 arrangement, called the *axoneme*, is at the heart of the movement of cilia, but although many speculations have been made, the fundamental mechanism is unknown. Cilia can be made to work isolated from the cells that give rise to them and, even in the absence of the ciliary membrane, by the addition of ATP under proper ionic conditions. There have been several suggestions that the beating of cilia is the consequence of a 'sliding-filament' motion between the components of the axoneme, but these suggestions on the whole have relied on arguments of analogy with muscle rather than direct evidence. There is, however, evidence of the involvement of Ca^{++} in the control process of ciliary beating, and this clearly strengthens the analogy.

The form of the beat of a typical cilium is shown in Figure 41. The total beat typically takes about 60 ms, about half in the effective stroke and half in the recovery stroke.

A protein called dynein, which has ATPase activity, has been isolated from the axoneme, and is thought to form the 'arms' that can be seen on the filaments in some cilia (Fig. 40). Although it is interesting that the ATPase protein should exist as a projection on a cylindrical structure, there have not been any other similarities reported between dynein and myosin, nor has there been any confirmed report of an actin-like protein from systems of cilia. For the time being, then, we do not seem to be able to extract from the motion of cilia any definite clues in our search for the fundamental mechano-chemical event.

Figure 41 The form of the beat of a typical cilium. The total time taken is about 60 ms.

Summary of Section 6.6.1

Microtubules are a class of cell components which are long straight hollow cylinders. Often they are arranged in special structures (cilia, etc.). When so arranged they may have an ATPase protein (dynein) projecting from the surface. Many roles, both structural and contractile, have been suggested for microtubules, but there is no clear evidence at the present time.

6.6.2   Bacterial flagella

For completeness, we must deal with the bacterial flagellum, while recognizing that it may be in a completely different class from those systems which we have discussed up to now. (Here the flagellum of a bacterium must be distinguished from the flagella of protozoan flagellates, which are in the same class as cilia, and have the standard 9 + 2 structure.)

bacterial flagella

protozoan flagella

Astbury called the bacterial flagellum a 'mono-molecular muscle', but this is an attribution that has never been proved. Pijper believes, after more than 30 years of study of cinematographic microscope films of bacterial motion, that the flagellum is a passive organ to give stability and that the bacterium itself moves, by undulating its body. Many people have rejected this view, and look upon the flagellum itself as the active agent in bacterial movement. So far this issue has not been resolved and it remains the most important problem in bacterial movement.

By analogy with muscle, a motile system might be expected to contain at least two proteins—the flagellum has one only (flagellin). An ATPase might be expected; none has been detected, nor is there any evidence for the presence of any phosphate compound like ATP. All these observations may make us doubt whether a flagellum can in fact transform chemical into mechanical energy, and may incline us to Pijper's view.

On the assumption that the flagellum is in fact active, a mono-molecular muscle as Astbury supposed, we may look at the structure of some flagella, and ask ourselves how they could change shape actively.

There have been many electron microscope pictures of bacterial flagella; typical ones are shown here (Fig. 42). The micrographs, together with X-ray diffraction studies, have been interpreted to show a close-packed helical surface array of globular sub-units of the protein. Models of these are shown in Figure 43, and are there compared with the actin helix. There is a superficial resemblance, in sub-unit size, but there is no doubt that the two proteins are distinct.

Figure 42 Electron micrographs of negatively-stained bacterial flagella ($\times 350\,000$)

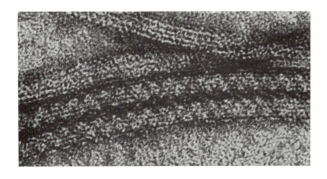
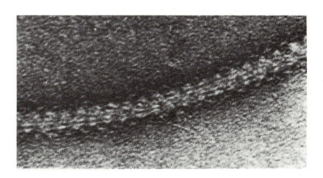

Figure 43 Ball models of the arrangement of sub-units in one type of bacterial flagellum.

Figure 44 The Spencer–Lowy mechanical model for one speculative hypothesis for flagellar movement.

(*a*) (*left*) The model consists of a large number of rings, each drilled with ten holes making an angle of 3° with the axis. Steel wires under tension join the rings together, with springs as spacers. The wires are terminated at the top on the circumference of a circular plate inclined to the horizontal; the highest point on the circle is transferred progressively round the rim of the plate by a motor-driven drive.

(*b*) (*right*) Close-up of part of the model, illustrating the fact that the wires run round the rod, making a small angle with the axis of the rod.

It would be out of place here to go in depth into the various modes of action that have been proposed for flagella because, as we have seen earlier, the fundamental question—whether the flagellum is an active or a passive organelle—is unresolved. However, we can at least mention one proposal. Spencer and Lowy point out that in a model such as that in Figure 43, a small change in shape in one helix along the direction of the dark balls would be amplified by the structure, and would throw the whole flagellum into a curled configuration. If the shape change were then repeated in a neighbouring helix the curled flagellum would seem to revolve. A mechanical model was built to demonstrate this hypothesis, and is shown in Figure 44. Certainly on this level it works, but it remains speculative, and it may not have much to do with bacterial motion.

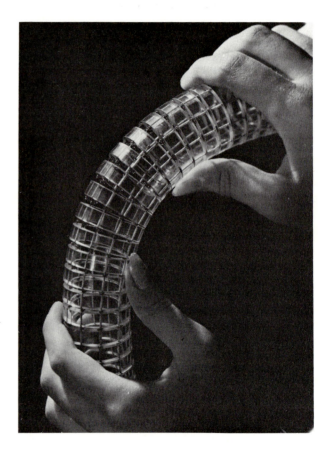

Summary of Section 6.6.2

It is not yet demonstrated that a bacterial flagellum is a 'mono-molecular muscle', but if so it seems it is unique in having no ATPase system. Perhaps the bacterium moves the flagellum, rather than the reverse.

Self-assessment questions for Sections 6.5–6.6.2

SAQ 21 (*Objective 3(a) and (c)*) In an obliquely-striated muscle, what structure is thought to be analogous to the Z-line in vertebrate striated muscle?

SAQ 22 (*Objective 3(a) and (b)*) Are the thin filaments in an obliquely-striated muscle similar in structure to those in vertebrate striated muscle?

SAQ 23 (*Objective 3(a) and (b)*) Are the thick filaments in an obliquely-striated muscle similar in structure to those in vertebrate striated muscle?

SAQ 24 (*Objective 3(a) and (e)*) What physiological explanation is there for the oblique striation?

SAQ 25 (*Objective 3(a) and (b)*) What evidence is there that the thick and thin filaments in a paramyosin smooth muscle slide past each other?

SAQ 26 (*Objective 3(a) and (e)*) Describe the 'catch' function of paramyosin smooth muscles.

SAQ 27 (*Objective 3(a) and (e)*) What evidence is there for a change in the arrangement of the molecules in the thick or thin filaments when a paramyosin smooth muscle goes into 'catch'?

SAQ 28 (*Objective 3(a) and (b)*) Are the thin filaments in a vertebrate smooth muscle similar in structure to those in a vertebrate striated muscle?

SAQ 29 (*Objective 3(b)*) What is the evidence for the occurrence of myosin filaments in a vertebrate smooth muscle?

SAQ 30 (*Objective 3(d)*) Draw a picture of the arrangement of 'sub-filaments', or microtubules, in a cilium.

SAQ 31 (*Objective 3(c) and (e)*) What is thought to be the location of the ATPase protein dynein in the diagram which you have drawn for SAQ 30?

SAQ 32 (*Objective 3(c) and (e)*) List three possible functions for microtubules.

SAQ 33 (*Objective 3(c) and (e)*) Does the bacterial flagellum move the bacterium, or the bacterium move the flagella?

SAQ 34 Astbury called a bacterial flagellum a mono-molecular muscle. Do you agree with him?

6.7 The characteristics of the contractile event

Our survey of muscular and non-muscular contractile systems in cells has shown very many diverse forms. To what extent can we draw general conclusions from the survey, to illuminate the basic mechano-chemical event?

Sadly, if any pattern emerges, it is at best a misty one, with large areas of uncertainty. Let us attempt to see some pattern. First, we can probably exclude bacterial flagella, for the reasons given at the beginning of Section 6.6.2.

Now try ITQs 3 and 4, and then go to the answers and remedial comments on p. 51 before you read the summary.

ITQ 3 Which items from the matrix below are characteristic of all the mobile systems discussed in Unit 6 (excluding bacterial flagella)?

(i) two sets of filaments	(ii) an ATPase protein	(iii) regularly arranged A-bands
(iv) filaments or similar structures are cylindrical (to a first approximation)	(v) characteristic actin-containing filaments	(vi) the interaction of two proteins in the contractile unit
(vii) regular side-by-side arrangement of filaments, at least in some regions	(viii) projections bearing the ATPase site	(ix) Ca^{++} as the initiator of activity
(x) a change of shape of the ATPase fragment	(xi) cyclic activity of the ATPase site	

ITQ 4 Which items from the same matrix are characteristic of all the muscle systems discussed in Unit 6?

Summary of the characteristics of the contractile event

All contractile systems probably have an ATPase protein, which seems to sit at the end of a projection from a cylindrical structure. These cylindrical structures tend to pack regularly, but do not always do so, it seems.

Excluding the non-muscle systems, there are always two major proteins, actin and myosin, contained in two sets of filaments, though the status of the thick filaments in vertebrate smooth muscle is still a little doubtful in living, resting muscle. In the non-muscular systems it has not been demonstrated clearly that motility demands the interaction of separate structural entities, although neither has the reverse been demonstrated.

The ATPase in the intact system is initiated by Ca^{++} and very probably acts cyclically in all systems. There is no clear demonstration of a change in shape of any ATPase fragment, which might cause the contractile force, but neither has there been a demonstration that such an event does not occur.

It is possible, though not proven, that there are two distinct types of contractile process in cells:

(a) using actin and myosin, for example, all muscles, slime moulds, blood-clot retraction, etc.;

(b) using microtubules, for example, cilia, mitotic spindle, etc.

A different explanation is also available for (b): this is that microtubules are skeletal elements which provide the framework for a (actomyosin?) contractile system.

6.8 How do contractile systems work?

Among research workers in this field there are, broadly speaking, majority and minority opinions about how contractile systems might work. It happens that we take a minority rather than the majority position. However, we have first acted as devil's advocate and put forward a model which intuitively we may not accept in all its aspects, even though it is accepted by very many workers in the field; we then offer you some thoughts about an alternative viewpoint.

We hope that you will realize that this is an area of current debate, that the weight which different workers put on different pieces of evidence is naturally not the same, and that we cannot always discuss the different interpretations at the length which they properly merit. We hope also that if your interest is aroused you will follow the matter further in some of the bibliography and references to this Unit. We would particularly draw your attention to Huxley (1969), Huxley (1973) and Offer (1973); these are very good accounts written from the majority viewpoint.

Finally, we ask you to note the form of Objective 4; specifically, we think that by the end of this Unit you will be able to discuss current ideas and theories of how contractile systems work, not that you will know the answers to all the questions which your discussion will arouse in your own mind.

6.8.1 The majority view

The active sliding force is developed by the heads of the myosin molecules which attach to the thin filaments and in doing so exert a longitudinal force on the thin filaments. These heads, the cross-bridges, act in a cyclical manner and pull the thin filaments along for a distance, probably 5.0 to 8.0 nm, and then they release and re-attach to the thin filament at another point. The cycle recurs and the total consequence is a continual motion of the thin filaments by the asynchronous action of all the cross-bridges acting on it from all the neighbouring, thick, myosin-containing filaments.

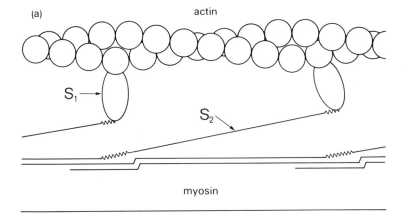

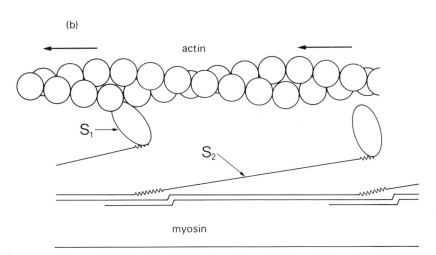

Figure 45 To illustrate the active change in angle of attachment of cross-bridges (Subfragment 1, shown on the diagram as S_1) to actin filaments could produce relative sliding movement between filaments maintained at constant lateral separation (for small changes in muscle length) by long range force balance. Bridges can act asynchronously as sub-unit and helical periodicities differ in the actin and myosin filaments. Only one of the two S_1 subfragments in each myosin molecule is shown.

(*a*) The left-hand bridge has just attached; the other bridge is already partly tilted.

(*b*) The left-hand bridge has just come to the end of its working stroke: the other bridge has already detached, and will probably not be able to attach to this actin filament again until further sliding brings helically arranged sites on actin into favourable orientation.

The mechanism requires the structural polarity of the thin and thick filaments which, as we have seen in Section 6.1.1, is a characteristic of the way in which these filaments are assembled. This is because the pull in one direction in one half of the A-band has to be opposite to the pull in the other half of the same A-band. In this way, the directions of the forces are specified by the structural polarities of the filament.

The relative force between actin and myosin is developed either as the result of an active change in the angle of attachment of the head of a myosin molecule to the actin-containing thin filament, or an active change in the shape of the myosin head, the attachment remaining fixed. This complex of the thin filament and the attached myosin head then is the basic contractile unit. This mechanism is shown in Figure 45.

It is not known whether the force-generating mechanism requires the co-operative attachment of both myosin heads at the same time to the thin filament, but it is thought possible that this may be so and that the contractile event may then be an (allosteric) shape or position change of the two heads acting together.

It is admitted that there is no technique which yet allows one to visualize the myosin head making its working stroke, but the consequence of the working stroke is observed in the way in which myosin heads attach to the actin filament and remain so attached under certain circumstances. This was shown in Figure 4 on p. 8 and is also shown in the results of H. E. Huxley's more recent image-reconstituting techniques; Figure 46 is a reconstruction of a decorated actin filament derived in this manner.

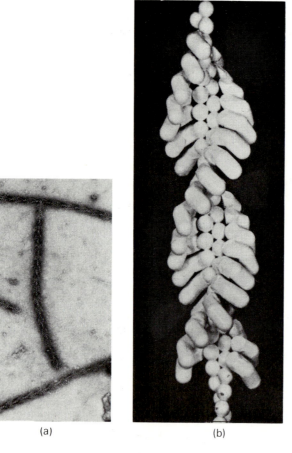

(a)   (b)

Figure 46(a) Electron micrograph of negatively stained preparation of actin filaments 'decorated' with myosin subfragment 1, showing well developed 'arrow-head' formations. (×100 000)

(*b*) Simplified model of 'decorated' actin filament, based on three-dimensional reconstruction results: the S_1 subfragments are attached to the central core of helically arranged actin monomers in a characteristically tilted and slewed configuration, and it is this that leads to the arrowhead appearance.

A final feature of the mechanism is that the attachment of the myosin heads to actin is controlled by the presence of calcium and, as we saw in Section 6.4.4, the tropomyosin–troponin system.

The attachment of actin to myosin is a necessary part of the generation of force by the actin–myosin system. The breakdown of ATP by the myosin alone is very slow, as we saw in Section 6.4.6, and this slow breakdown develops no tension in the absence of actin–myosin head attachment. Under these conditions, the muscle can be pulled out, with the filaments sliding past each other by a relatively small external force.

One of the advantages of any hypothesis depending on the action of cross-bridges is that the cycling of the cross-bridges can be compared with the steps in the Lymm–Taylor scheme for ATP hydrolysis (this is discussed in Section 6.4.6) in a slightly more complicated form to take account of the effect of actin. With the additional assumption of some elasticity in the flexible link of the cross-bridge it is possible to make predictions about the mechanical behaviour of the muscle fibre, and the changes in the orientation of the head at various positions in the cycle. The transient mechanical effects seem to be consistent with this model, though they might also be consistent with other models. The X-ray technique is unfortunately not yet sensitive enough to detect active changes in head orientation during the different phases of the cycle, though several laboratories are trying to develop much more powerful X-ray sources, or detection procedures, in an effort to break through this barrier.

In a mechanism such as that shown in Figure 45 on p. 43, the change in the shape of the myosin head produces a relative force between actin and myosin, but also must produce a relative transverse force at right angles to the actin and myosin filaments. It is a necessary part of the mechanism that this transverse force is absorbed in the stability of the system, absorbed, if you like, in the force balance which we talked about in Section 6.3.

To summarize, according to the majority view, muscle contraction is caused by a rather precise structural event taking place in the cross-bridge at the point of myosin-head attachment to actin and the forces produced by these structural events are added together because of the structural polarity of the filaments. The total force is further smoothed out by the asynchronous motion of the cross-bridges to give the macroscopic muscle tension.

Huxley (1973) has shown that such a mechanism can plausibly account for the contraction of many types of muscle apart from striated muscle and also can be made to account for the contraction of some of the non-muscular systems (particularly cytoplasmic streaming and cell movement) by postulating an active shearing force developed between two sets of filaments together with the attachment of some of the filaments to the surface of the cell. Huxley's (1973) paper is well worth reading as a follow-up to this Unit.

6.8.2   A minority view

The diversity of forms in which contractility and motility are possible can be thought to suggest a rather generalized overall reaction rather than a precise structural event.

One minority opinion is that the small changes which have been seen in the structures are the consequences of local changes in the electrical fields rather than of actual mechanical strains providing a driving force. Thus, as we described under the majority viewpoint, we could consider that a myosin cross-bridge, moving toward an actin filament, does so to attach and deform, pushing the actin along, *or* it may do so (as you read in Section 6.3) because the local arrival of the positively-charged calcium ion on to a specific site on the actin filament changes the equilibrium position of the negatively-charged myosin head with its attached ATP, also negatively charged. This change in local electrical conditions could be the switch-on mechanism for the enzyme. In support of this view, it is noticeable that in many enzyme systems far removed from muscle the active site is held near to a charged surface, and in the region of high potential gradient (mitochondrial membranes come to mind).

You saw in Section 6.3 that relatively small changes in the electrical conditions, for example in the pH and ionic strength, can cause the whole muscle lattice to swell or shrink. The forces involved in this swelling and shrinkage can be considerable, but of course this is a lateral swelling and shrinkage rather than a longitudinal force. The problem in generating some mechanism for muscle contraction which depends on this swelling and shrinkage is not that of producing the swelling and shrinkage itself. This is not a difficult problem because it is easy to postulate mechanisms of enzyme activity which might change the local charges, and therefore the local potentials in the appropriate way. It is not so easy, though, to devise a mechanism where this swelling, transversely oriented, is converted into longitudinal force. One possibility is to use the structural polarity of the filaments and the attachment of the myosin cross-bridges to the thin filaments. On this view they are used in a more passive way than is envisaged in the majority viewpoint. That is, we may suppose (Fig. 47) that the transverse swelling is transmitted to the other filament by the action of the attached cross-bridge, which now behaves as a dragging anchor. When its orientation is no longer appropriate for transmitting force, it drops off and re-establishes itself at a different length.

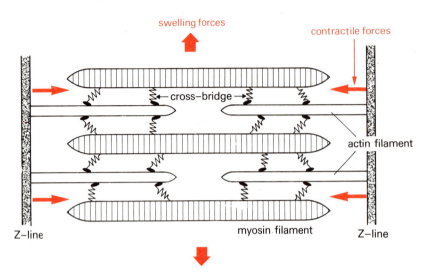

Figure 47 An alternative mechanism. Swelling forces developed in a A-band are transmitted to the filaments by the myosin heads acting as anchors. When the flexible link (the anchor chain) no longer is in the appropriate direction to transmit tension the anchor drops off and re-establishes itself a new position.

This mechanism, like the majority one, involves some cyclic activity of the attached myosin cross-bridges, just as it requires the observed polarities of the thick and thin filaments. It has been known for more than 20 years (Bozler, 1951) that if a (glycerinated) muscle in rigor (which means that the cross-bridges are attached) is put into a solution of low ionic strength the muscle will develop tension, but will not shorten. Rome's results (see Fig. 15 on p. 17) show that such a muscle will swell, so that the observation is consistent with the hypothesis given above, in the special condition where the cross-bridges cannot detach and form again at a different length.

Appendix 1 (black-page)   X-ray diffraction in biology

We shall not expect you, after reading this Appendix, to be able to interpret X-ray diffraction diagrams from biological fibres. This, as we said in the main text, is a highly technical matter and is in no sense part of our Course Objectives. We will try, though, to give you some feeling for the sort of information that is available from X-ray diffraction of biological fibres and for the sort of circumstances under which these methods may be used.

You met diffraction in S100[5] and you may remember the general phenomenon—electromagnetic radiation hitting matter of any sort is scattered in all directions by that matter, but if the scattering matter contains some sort of periodic array, then the interference of the various secondary wavelets that are scattered may be additive in specific directions so that diffracted beams are set up in these particular directions. You saw this in the diffraction grating which some of you used in the S100 Summer School and in the demonstrations which were given in the TV programme for Unit 28 of S100. (You can if you wish refer to the TV Notes for this programme.)

The applications of diffraction which we have just mentioned were of the diffraction of *light* by periodic structures. The periodic structures were either one-dimensional (the diffraction grating) or two-dimensional (the periodic grids used in S100, TV programme 28). The periodic dimensions of these structures were all of the order of a few micrometres, this was governed by the wavelength of the light which was of the order of 0.5 μm.

X-ray diffraction

The principle of X-ray diffraction is precisely the same as that of light diffraction, except that the wavelength of the electromagnetic radiation is now different and is of the order of 0.1 nm. The size of the regularities which then give rise to the diffraction effects are of the order or 0.1–100 nm and these sorts of periodicities are concerned with the spacings of atoms in crystals, or possibly the spacings of molecules, or filaments made of molecules, in biological tissues. The assemblies which make up the diffracting structure are now usually three-dimensional rather than two-dimensional (you may think of the atoms of sodium and chlorine in a crystal of sodium chloride) but this does not cause any difficulty in principle, although of course it makes the mathematics a little more complicated.

Several of the Open University Second Level Science Courses have included discussions of X-ray diffraction, so that you may already have met the use of X-ray diffraction as a method of finding out positions of the atoms of a chemical compound.[11] Some of the applications of X-ray diffraction to biology have used this concept and have produced very elegant visualizations of the structure of biological crystals containing many hundreds of atoms. A particular example of this was the three-dimensional structure of an enzyme, in the lysozyme, by D. C. Phillips and his collaborators (Phillips, 1966).

In work of this kind the principles which are used are exactly the same as the original principles used by the pioneers to find out the structure of simple crystals like sodium chloride and, although the technical complications of the process go up markedly with the number of atoms in the molecule, there are few if any differences in principle. But in this Appendix we shall not deal with this aspect of X-ray diffraction applied to biology, but rather shall talk about a slightly different application of the technique, to the diffraction from fibrous structures.

It is rare in a biological fibre for enough regularity to be present in three dimensions for the position of each atom to be determined by the X-ray diffraction technique, although this has been possible in the special case of the fibre of DNA. More often, however, one cannot find out much information about the arrangement in individual atoms in a biological fibre, but one can find out information about the arrangement of molecules or assemblies of molecules; in muscle we can think of the assembly of the myosin molecules in the thick filament or the assembly of the actin molecules in the thin filament. The periodicities, or spacings, involved here are at least in order of magnitude greater than the distances between the actual atoms, say from 1–10 nm.

You may recall[5] that the grating equation is of the form $d \sin \theta = n\lambda$, which means that the longer is the diffracting spacing (d) the smaller will be the deviation (θ) of the diffracted beam from the main beam. To observe spacings of 10 or more nanometres, it is necessary to use technically advanced methods which will enable the experimenter to measure very small angles of diffraction. This is a problem which has been solved over the past 20 years, and we shall not spend time in discussing the technical details. These low-angle (referring to the angle of deviation) diffraction pictures are often used as 'fingerprints' of biological structures. If you look in more detail at Figure 18 in the main text (which is repeated here as Fig. 48) you will see that there are a series of X-ray

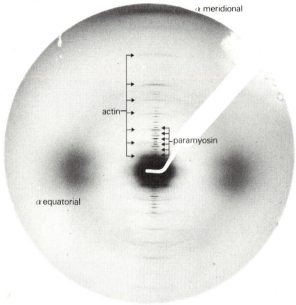

Figure 48 The X-ray diagram from obliquely-striated oyster adductor muscle, which is rich in actin-containing filaments. We have marked the diffuse actin reflections, the sharper paramyosin reflections and the 'α' reflections from the molecular structure of the helices.

reflections (which represent beams diffracted by the fibre) which we can associate with the paramyosin filaments, and there is indeed a third series which are associated with the atomic structure within the myosin filaments. It is convenient to define some directions in the X-ray pattern. The usual terminology is to talk about the 'meridian' of the pattern, that is, beams which are diffracted in the same direction as the biological fibre itself and the 'equator' of the pattern, that is, beams which are diffracted at right angles to the fibre. This is shown in Figure 49. Returning to Figure 48 we will look first of all at the series of reflections which are furthest away from the main beam. These are a very strong equatorial reflection at about 0.9 nm and a strong meridional reflection at about 0.15 nm. This is the characteristic 'alpha' pattern observed by Astbury many years ago from a wide variety of biological fibres and associated with the Pauling–Corey alpha helix. In atomic terms this is not a vey detailed X-ray pattern and does not give us a great deal of information about the atomic structure of the alpha helix, although it does fit in with all the many other pieces of chemical evidence which Pauling used in his original hypothesis.

Figure 49 The terminology used in fibre X-ray diagrams.

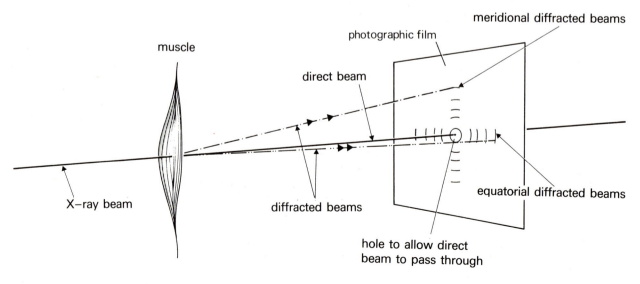

Next, consider the series of rather diffuse reflections associated with the actin filaments. Even to the untutored eye, they appear different from the alpha reflections we have just talked about and from the much sharper series of reflections associated with the paramyosin filaments. The actin reflections, as we have implied in the main text, can be taken as a fingerprint for the occurrence of actin filaments. The diffuseness of the X-ray reflections tells us something about the size of the filaments in which these molecules occur and because the actin reflections are diffuse we know that the actin filaments are likely to be smaller in size than the filaments which give rise to the second, sharper series of reflections associated with paramyosin. By making detailed measurements of these two series of reflections in the meridional direction, we can work out the basic spacings in the actin and in the paramyosin filaments, much as you worked out grating spacings with light diffraction if you did the Summer School experiment for S100. We can make a list of these spacings for various fibres of biological interest (Table 2) but, as is apparent from Figure 48, it is not just

Table 2   X-ray measurements of filament-axis spacings

Filaments	Spacing/nm
Myosin	42.9 (3 × 14.3)
Paramyosin	72.0 (5 × 14.4)
Actin	36.0
Collagen	64.0–67.0

the spacing measurements which distinguish between the series of reflections characteristic of actin and the series of reflections characteristic of paramyosin; the paramyosin reflections are much sharper and are easily distinguishable in the X-ray diagram. In this sort of way scientists learn to recognize the fingerprints of different sorts of fibres in biological tissue. Notice that the X-ray method is applicable to living intact tissue; one needs only to dissect a muscle or a nerve from an animal and to place it in a suitable Ringer solution and pass a beam of X-ray through it—the Ringer solution will maintain the muscle or nerve fibre in a living condition and we can test the physiological condition by stimulating the muscle and seeing if it will still contract or by stimulating the nerve and seeing if it will still conduct an impulse. In this way we can get structural information from a piece of tissue which has not been subjected to the stringent procedures which are necessary, for example, for examination in the electron microscope (see Unit 1).

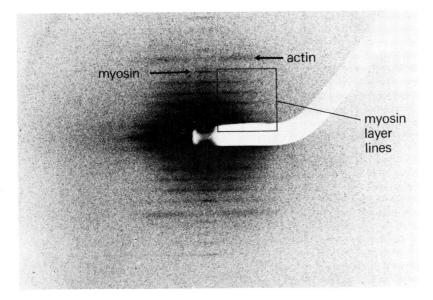

Figure 50 The X-ray diagram from frog striated muscle, showing the meridional myosin reflections, the 5.9 nm actin reflections and the layer-lines which arise from the projections on the myosin filaments (the cross-bridges).

Figure 50 is the X-ray diagram from frog striated muscle and is at a larger magnification than Figure 48 so that the reflections which you see come from longer spacings than those in Figure 48 (remember the reciprocal relationship). The diffuse reflections marked actin are at a meridional spacing of about 5.9 nm and the major meridional reflections of myosin are at 14.4 and 7.2 nm. The most interesting feature of Figure 50 is the pattern which occurs in the quadrants of the picture; these are called the myosin layer-lines. This pattern arises because of the very regular helical arrangement of cross-bridges around the shaft of the myosin filament. It is this pattern which becomes very much less intense when the muscle contracts, showing that the regular arrangement of the cross-bridges is disturbed and implying that they are taking part in some asynchronous motion because no new regular pattern appears in its place in contracting muscle.

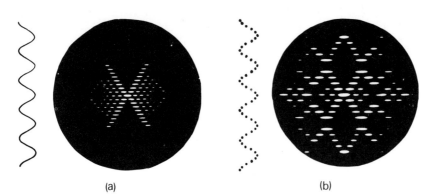

(a)   (b)   (c)

It may be of interest to you to look at Figure 51(a)–(c) which shows the light analogue for helical diffraction; the diffraction patterns are obtained from the masks shown by using a laser light and a photographic plate. Notice that the major feature of helical diffraction is a diamond shaped region. This will also have been apparent to you if you watched the TV programme for S100, Unit 28 and you can see in the TV notes for that programme the X-ray diagram of the DNA molecule compared with the light diffraction from a series of regular points on a helix.

Figure 51(a)–(c) The light analogue for helical diffraction. Notice that the major feature is a diamond shaped region free of diffraction, and that the major diffraction is neither on the meridian nor on the equator but is in the quadrants of the diagram (cf. the myosin layer lines, Fig. 3).

Equatorial diffraction

So far we have talked about meridional X-ray diffraction arising from the backbone structure of filaments, or the arrangement of projections round actin and myosin filaments. These occur in the meridional or fibre axis direction of our X-ray diagram. In most muscles, however, the strongest X-ray reflections come not from the backbone structure of the filaments, but from the packing of the filaments among themselves. The most regular thing in striated muscle is the packing of actin and myosin filaments in the regular hexagonal array which you saw in Figure 5.1 of Huxley's article. So strong are the X-ray reflections which arise from this packing that in Figure 50 the equatorial region of the pattern is grossly over-exposed and does not show the regular reflections. These are shown in Figure 41, which is presented in the same orientation so that you can think of the muscle fibre as running from top to bottom of the page. In Figure 52 you see clearly the two equatorial reflections which are characteristic of vertebrate striated muscle and which arise from the packing of the filaments into the hexagonal lattice. We are not going to talk about this in detail here because this point will be taken up in the TV programme for Unit 6 where we hope to show you the sort of information that we can get from these two reflections and the inferences which we can make about the structure of muscle and about its physiological functioning.

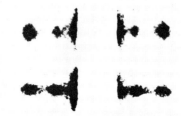

Figure 52 The equatorial pattern from vertebrate striated muscle, showing the two strong reflections which arise from the packing of the filaments into the hexagonal lattice (see also the TV programme for Unit 6).

In that programme we shall stress the two types of information which are available from X-ray diffraction:

1 Information of *position*, which gives the dimensions of the diffracting structure.

2 Information from the *intensities* of the X-ray reflections, which can in principle tell us something of the way in which matter is arranged in the diffracting structure.

X-ray diffraction as a control procedure

Finally in this Appendix it seems appropriate to stress once more the point made a few paragraphs ago. X-ray diffraction is a technique which can be applied to functioning physiological active tissue; it requires no preparation procedures which involve killing the tissue, fixing it, sectioning it and so on. In those tissues where the regularity is sufficient for X-ray diffraction to be effective, this is an extremely powerful technique for the elucidation of biological structure in the *in vivo* condition. It is also a powerful control procedure for the results obtained from other more indirect methods: in muscle, for biochemical investigations of the muscle proteins and the sorts of filaments which they form *in vitro* and for the interpretation of the electron micrographs which have been obtained from very many muscles. Indeed, it is true to say that some of the interpretations of the electron micrographs would have led to possibilities of error had the X-ray evidence not also been available.

Answers to ITQs

Notice that some of these answers are matters of opinion, and you may have good reasons for dissenting from our opinion. This is what University education is all about!

ITQ 1 (iv), (vi) and (xii)
We think these are the *essential* features.

(i), (iii), (viii) and (x)
These are important features of the evidence which first led to the proposal of a sliding-filament model. They are not, however, essential features of the model itself.

(ii) and (vii)
These are interesting facts about myosin, but again, not essential features of the model in this context.

(v), (ix), (xi), (xiii), (xiv) and (xv)
Interesting facts about muscle as a whole but not essential features of the model in this context.

ITQ 2 (a) (ii), (iii), (vii), (ix), (xi) and (xiii)
These seem to be best in the context of this question. Item (ii) is

included, though perhaps should be given less weight than the other items.

Items (i), (iv), (vi), (viii) and (xii) are already defined by the terms of the question, and not appropriate as responses to that question. (See the answer to ITQ 1.)

(b) The remaining items, (v), (x), (xiv) and (xv) are, in our opinion, appropriate responses to question 2(b).

ITQ 3 (ii), (iv), [(viii)], (ix) and (xi)
You may not be very impressed by this list of characteristics, and may even wonder whether the non-muscle systems have a totally different contractile system to the muscle ones. The question is open. We have included (viii) in parenthesis, because for some systems (microtubules) this may not apply, though it seems to when the same systems are built into cilia, etc.

ITQ 4 (i), (ii), (iv), (v), (vi), (vii), (viii), (ix) and (xi)
Here the list looks more impressive. We have not included item (x). Many workers might do so, but on balance we do not believe that this has been demonstrated in any unequivocal way.

Answers to SAQs

SAQ 1 (a) (i); (b) (ii); (c) (v); (d) (ix); (e) (xii).
These should all be clear from Huxley's article in *BBB* and the postscript in Section 6.1.1.

SAQ 2 (a) Your diagram should show the two twisted rows of globules. See Figure 3(b).
(b) Your diagram should show the bare zone, and the polarity of the myosin heads projecting from the filament. See Figure 2.

SAQ 3 (ii) is correct, see the discussion in Section 6.1.2.

SAQ 4 Your diagram should look like Figure 11, and your account should contain the major points made in Section 6.1.4.

SAQ 5 (ii) is correct. See Figure 10.

SAQ 6 (iii) is correct, see the discussion in Section 6.2.

SAQ 7 (iii) is correct, see the discussion in Sections 6.2 and 6.4.4.

SAQ 8 No. Force is produced only during contraction; relaxation is a passive process. (See Section 6.1.1.)

SAQ 9 (iv) and (vi). See Section 6.4.4.

SAQ 10 (i) and (vii). See Section 6.4.1.

SAQ 11 (iii) is not an alpha protein in the sense defined, although it very likely does contain some alpha-helical chains. (ii), (iv) and (v) have greatest helix content. See Table 1.

SAQ 12 (iii). See Section 6.4.2.

SAQ 13 (iv) and (vi). See Sections 6.4.3 and 6.4.4.

SAQ 14 (v). See Section 6.4.5.

SAQ 15 (iii) and (vii). (A possible alternative answer would be (iii) and (i).) This is discussed implicitly throughout this Unit. See Section 6.4.4 in particular.

SAQ 16 (vii). See Table 1.

SAQ 17 Because it shows that the two major proteins must interact over a surface-to-surface distance which varies by a factor of 2.5 in the physiological range.

SAQ 18 Because it shows that the myosin projections, or 'cross-bridges' which give rise to this pattern, must move around, become disordered or in any event undergo some sort of displacement during the contraction.

SAQ 19 No, they do not. See Section 6.1.1.

SAQ 20 Inversely related. The higher the load, the slower the speed until at some high load the muscle can no longer shorten at all. See Section 6.1.3.

SAQ 21 The dense bodies. See Section 6.5.2.

SAQ 22 Yes. See Sections 6.4.2 and 6.5.2.

SAQ 23 No, they are paramyosin filaments. See Sections 6.4.5 and 6.5.2. But notice that, as we said in Section 6.5.2, paramyosin filaments have a surface coating of myosin molecules.

SAQ 24 No convincing explanation has yet been offered for this.

SAQ 25 The evidence is of two main types (Section 6.5.3):
(a) The X-ray results show that there is no major molecular reorganization as the muscle length changes.
(b) Filament counts show that there are more filaments in the *total* cross-section of a short cell compared with a large one.

SAQ 26 Many paramyosin smooth muscles can remain contracted for long periods, with little or no expenditure of metabolic energy. This is a very desirable response in a muscle which may be used to hold the two halves of the animal's shell together against a predator, or to anchor the whole animal on to a rock.

SAQ 27 There is no good evidence for this, and very strong X-ray evidence that the arrangement does *not* change in either set of filaments.

SAQ 28 Yes. See Sections 6.4.2 and 6.5.4.

SAQ 29 The strongest single piece of evidence is the meridional X-ray reflection at 14.4 nm. There is also electron-microscope evidence for a second, thicker set of filaments in some (though not all) circumstances.

SAQ 30 Your diagram should show the 9 + 2 arrangement and the arms on the outer tubules (see Fig. 40 on p. 38).

SAQ 31 At the ends of the arms on the outer tubules (see Fig. 40).

SAQ 32 These are listed in Section 6.6.1.

SAQ 33 We do not know the answer to this question, though many scientists who work with bacterial flagella assume that the flagellum is the active agent. This is the subject of some hydrodynamic controversy.

SAQ 34 An open-ended question. When Astbury made this speculation, three decades ago, he believed that muscular contraction involved the deformation of long myosin molecules into another, more twisted form. We now know that this does not happen. Of course, as we do not yet know for sure what is the underlying physical event in either muscle contraction or (bacterial) flagellar motility, it may yet turn out that there are similarities which justify Astbury's remark.

Bibliography and references

We hope that at least some of you may want to go further with the concepts we have met so far, or with some, like the energy requirements, which we have only hastily touched upon. So here are a few suggestions for further reading.

1 General

Wilkie, D. R. (1968) *Muscle*, Edward Arnold. 43p.
(A short account (65 pp.), very readable, particularly useful on control, energy balance and power production. No references (or very few).)

Bendall, J. R. (1969) *Muscle, Molecules and Movement*, Heinemann.
(A longer, and comprehensive, scientific monograph, with full references to the journals. Written in 1969 but still reasonably up-to-date. A good starting point for an in-depth study.)

Ebashi, S. (1972) 'Calcium ions and muscle contraction', *Nature*, **240**, 217–18.
(Two pages which summarize the calcium, trypomyosin–troponin story, and give full references.)

Elliott, G. F. (1973) 'The muscle fiber: liquid-crystalline and hydraulic aspects', *Ann. N.Y. Acad. Sci.*, **204**, 564–73.
(This ten-page article expands on some of the concepts in Sections 6.3 and 6.8. It was written as a review article, in a reasonably non-jargon genre.)

Huxley, H. E. (1965) 'The mechanism of molecular contraction', *Sci. Amer.*
(*Scientific American*, Reprint No. 1026, 11p. by post from W. H. Freeman & Co.)
(An expansion of some of the concepts in this Unit, which could usefully be read in conjunction in Section 6.1.1.)

Huxley, H. E. (1969) 'The mechanism of muscular contraction', *Science*, **164**, 1356.
(H. E. Huxley's fullest statement of his view of the contractile process. Gives very comprehensive references.)

Offer, G. (1973) 'The molecular basis of muscular contraction', *Companion to Biochemistry*, Chapt. 22, eds. Bull, A., Lagnado, J., Thomas, J. and Tupton, K., Longmans. (In press.)
(A very good, and very up-to-date account, written for undergraduate readers. A few well chosen references.)

Porter, K. R. and Franzini-Armstrong, C. (1965) 'The sarcoplasmic reticulum', *Sci. Amer.* (March 1965) (*Scientific American* Reprint No. 1007).
(A useful expansion of Sections 6.2 and 6.4.4, showing how the sarcoplasmic reticulum performs the functions of rapid telegraphy, plumbing and supply.)

Holmes, K. C. and Blow, D. M. (1966) *The Use of X-ray Diffraction in the Study of Protein and Nuclear Acid Structure*, Interscience.
(In case you want to go further with the techniques and concepts in Appendix 1.)

2 References cited in the text

Bozler, E. (1951) 'The effect of electrolytes on the contractile elements of muscle', *J. Gen. Physiol*, **35**, 703.

Ebashi, S. (1972) 'Calcium ions and muscle contraction', *Nature*, **240**, 217–18.

Elliott, G. F. (1973) 'The muscle fiber: liquid-crystalline and hydraulic aspects', *Ann. N.Y. Acad. Sci.*, **204**, 564–73.

Huxley, H. E. (1965) 'The mechanism of muscular contraction', *Sci. Amer.* (Dec. 1965). (*Scientific American* Reprint No. 1026).

Huxley, H. E. (1969) 'The mechanism of muscular contraction', *Science*, **164**, 1356.

Huxley, H. E. (1973) 'Muscular contraction and cell motility', *Nature*, **243**, 445.

Matsubara, I. and Elliott, G. F. (1972) 'X-ray diffraction studies of skinned single fibres of frog skeletal muscle', *J. Mol. Biol.*, **72**, 657–69.

Offer, G. (1973) *Companion to Biochemistry*, Chapt. 22, eds. Bull A., Lagnado, J. and Tipton, K., Longmans.

Phillips, D. C. (1966) 'The three-dimensional structure of an enzyme molecule', *Sci. Amer.* (Nov. 1966). (*Scientific American* Reprint No. 1055.)

3 References to material in prerequisite and other science courses

1 S100, Unit 18
2 S100, Unit 18
3 S100, Units 3 and 4
4 S100, Unit 5
5 S100, Unit 28
6 S100, Unit 2
7 SDT 286, Unit 1
8 S100, Unit 5
9 S2–1, Unit 1
10 S2–1, Unit 1, Fig. 19
11 S24–, Unit 2

Acknowledgements

Grateful acknowledgement is made to the following sources for material used in this Unit:

Figure 1 Susan Lowey and Henry Slayter; *Figures 3a and 3b* Jean Hanson and Jack Lowy; *Figures 4a, 4b, 13 and 14* Heinemann Educational Books Ltd for H. E. Huxley and J. R. Bendall, *Muscles, Molecules and Movement*; *Figure 5a* G. F. Elliott, J. Lowy and B. M. Millman (adapted from Fig. 4, pp. 31–45, *J. Mol. Biol.*, **25**, 1967); *Figure 11* The Physiological Society for A. M. Gordon, A. F. Huxley and F. J. Julian in *J. Physiol.*, **184**, 1966; *Figure 12* from 'The sarcoplasmic reticulum' by K. R. Porter and C. Franzini-Armstrong, Copyright © 1965 by Scientific American, Inc. All rights reserved; *Figure 15* New York Academy of Sciences for Elizabeth Rome in G. F. Elliott, pp. 564–74, *Ann. N.Y. Acad. Sci.*, **204**; *Figure 18 (repeated as 48)* B. M. Millman and G. F. Elliott; *Figure 19* J. M. Gillis, 'Le role du Calcium dans la controle intracellulaire de la contraction musculaire', Vander/Louvain (ed) 1972; *Figures 20, 22, 34, 45 and 46* Nature and the authors for S. Ebashi in *Nature*, **240**, 218, 1972, Jean Hanson and Jack Lowy in *Nature*, **180**, 906, 1957, J. Lowy in *Nature*, **225**, 1053, 1970, and H. E. Huxley in *Nature*, 243, 1973; *Figures 21, 35 and 38* The Royal Society for G. F. Elliott, p. 468, C. F. Shoenberg and D. M. Needham, p. 523, and Jean Hanson and J. Lowy, p. 524 all in *Proc. Royal Soc.*, B, **160**, 1964; *Figures 24 and 25* J. Rosenbluth in *J. Cell. Biol.*, **25**, 495, 1965; *Figure 26* Jean Hanson and J. Lowy; *Figure 27* C. Franzini-Armstrong; *Figures 28, 29 and 42* Jean Hanson and J. Lowy; *Figures 30 and 39* Academic Press, Inc. (London) Ltd. for G. F. Elliott in *J. Mol. Biol.*, **10**, 89, 1964 and V. Nachmais and H. E. Huxley in *J. Mol. Biol.*, **150**, 83, 1970; *Figure 32* Rockefeller University Press for B. M. Millman and G. F. Elliott in *Biophys. J.*, **12**, 1409, 1972; *Figure 33* Meredith Publishing Company for C. F. Shoenberg in R. W. Wynn (ed), *Cellular Contraction of the Uterus* published by the Appleton-Century-Crofts Division; *Figure 36* R. V. Rice; *Figure 37* A. P. Somlyo and R. V. Rice; *Figure 40* Academic Press, Inc. for B. A. Afzelius in *J. Ultrastruct. Res.*, **37**, 195, 1971; *Figure 43* J. Lowy and Jean Hanson in *Nature*, **202**, 538–40, 1964; *Figure 44* Cambridge University Press for J. Lowy and M. Spencer in *Aspects of Cell Mobility*, 1968; *Figure 50* G. F. Elliott, J. Lowy and B. M. Millman in *J. Mol. Biol.*, **25**, 31–45, 1967; *Figure 51* John Wiley and Sons, Inc. for K. C. Holmes and D. M. Blow in D. Glick (ed.), *Methods of Biochemical Analysis*, **13**, 113–239.

Unit 7

Cell Growth and the Cell Cycle

Contents

Table A: part 1

Principal Sections of recommended prerequisite Courses on which this Unit depends

Course	Unit and Section No.	Topic
S100*	14.4	form and function of cells
	14.5	ultrastructure of cells
	17.2–17.5	DNA and RNA structure
	17.6–17.9	transcription and translation
	17.11–17.14	cell division; the cell cycle
S2–1**	6.4	enzyme induction
	6.6	the control of DNA synthesis and cell division
S2–5***	1.2	basic genetics
	2.4, 2.5	protein synthesis, enzyme induction
	3.2	differential gene transcription
	3.5	temporal control

* The Open University (1971) S100 *Science: A Foundation Course*, The Open University Press.

** The Open University (1972) S2–1 *Biochemistry*, The Open University Press.

*** The Open University (1973) S2–5 *Genes and Development*, The Open University Press.

Table A: part 2

List of scientific terms, concepts and principles used in Unit 7

Developed in this Unit	Page No.	Developed in a later Unit	Unit No.
balanced growth	9		
chromosome segregation	21		
Cooper–Helmstetter model	19		
division cycle, DNA cycle	29		
division proteins	26		
Gl, S, G2, M	23		
growth cycle	29	biological clocks	15
immunofluorescence	14		
induction synchrony	7		
initiation of DNA synthesis	21		
linear and exponential growth increase	12		
linear transcription model	16		
oscillatory repression model	15		
periodic and continuous protein synthesis	14		
pulse labelling	13		
selection synchrony	8		
synchronous growth	10		

Objectives for Unit 7

When you have completed this Unit, you should be able to:

1 Demonstrate an understanding of the terms, concepts and principles listed in Table A.
(SAQs 1 and 19)

2 Identify characteristics of G1, S and G2 and calculate their duration.
(SAQ 15)

3 Distinguish between methods of induction and selection synchrony and explain their respective advantages and disadvantages.
(SAQs 2–4)

4 Name different criteria for measuring cell growth and explain why measurement of growth should involve the use of more than one criterion.
(SAQ 5)

5 Discuss the growth of two named organelles during the cell cycle.
(SAQs 6 and 7)

6 Describe and evaluate the evidence for the oscillatory repression model and for the linear transcription model for periodic enzyme synthesis.
(SAQs 8 and 9)

7 Describe the Cooper–Helmstetter model for the integration of DNA synthesis in the bacterial cell cycle.
(SAQs 10, 12 and 13)

8 Compare and contrast DNA synthesis in procaryotes and eucaryotes.
(SAQ 14)

9 Evaluate evidence from cell fusion and nuclear transplantation experiments relating to the initiation of DNA synthesis in eucaryotes.
(SAQ 16)

10 Discuss the evidence for 'division proteins'.

11 Briefly describe the evidence for the different types of cycles within the mitotic cell cycle of eucaryotes.
(SAQ 11)

12 Define the terms 'execution point' and 'termination point' and explain the experimental results that led to these concepts.
(SAQ 17)

13 Explain why repeated heat shocks in *Tetrahymena* induce synchronous growth.
(SAQ 18)

7.0 Introduction and study guide to Unit 7

7.0.1 Studying Unit 7

This Unit concludes the first part of the Course which is concerned with various aspects of cellular physiology. It discusses how the organized structure and function of the cell described in earlier Units is developed and maintained throughout the cell cycle. Because these cellular processes are regulated primarily by events within the nucleus, our attention is focused on the replication of DNA and the regulated expression of genes. As the Unit emphasizes the biochemical and genetical aspects of cell physiology, you will need to have studied both S2–1 *Biochemistry* and more particularly S2–5 *Genes and Development*.

Three (black-page) papers in your offprint collection relate to this Unit. These are:

Cooper, S. and Helmstetter, C. E. (1968) 'Chromosome replication and the division cycle of *Escherichia coli* B/r', *J. Mol. Biol.*, **31**, pp. 519–40.

Tauro, P., Halvorson, H. L. and Epstein, R. L. (1968) 'Time of gene expression in relation to centromere distance during the cell cycle of *Saccharomyces cerevisiae*', *Proc. Natl. Acad. Sci. U.S.A.* **59**, pp. 277–84.

Hartwell, L. H., Culotti, J. and Reid, B. J. (1970) 'Genetic control of the cell-division cycle in yeast. I. Detection of mutants', *Proc. Natl. Acad. Sci. U.S.A.* **66**, (2), pp. 352–9.

If you are interested and have time, you should study Tauro's paper while you read Section 7.2.4(b) and Hartwell's paper while you read Section 7.5.2.

When you read Section 7.3.1 you may have problems understanding the Cooper–Helmstetter model of DNA replication, which is an interesting but difficult concept to grasp. If you cannot appreciate the model and are short of time, pass on to the remainder of the Unit. If you have time and are interested, you can read the paper by Cooper and Helmstetter which discusses the model, and the experimental evidence for it, in some detail.

7.0.2 Introducing the cell cycle

The cell cycle of most cells is made up of two physiologically distinct processes. During *interphase*, which occupies about 95 per cent of the cycle, the cell grows and the components of the cell increase. *Mitosis*, which is relatively short, ensures the division of the enlarged cell into two daughter cells so that the products of interphase are apportioned equally.

interphase
mitosis

We know that growth as such is not necessary for cell division because embryos can divide without concomitant growth.[1] However, it is likely that certain metabolic events take place during interphase, linked in a sequence which results in mitosis. During interphase there must be some metabolic activities which are concerned with growth of the cell, but also there may be activities, such as DNA synthesis, which are involved in the control of cell division.

Until recently very little was known about interphase. Our knowledge was greatly enhanced when Howard and Pelc (1953) discovered, using autoradiography, that radioactively labelled precursors of DNA were incorporated into DNA only during a discrete part of interphase. They divided the interval between mitosis, that is interphase, into three phases. The G1 phase followed mitosis and preceded DNA synthesis. The S phase was the period of DNA synthesis and the G2 phase separated the S phase and mitosis, M. The four phases, G1, S, G2 and M, make up the cell cycle of eucaryotes (see Fig. 1).

Within the mammalian body, some cells divide repeatedly, for instance cells in the lining epithelium of the crypts of the small intestine. Others, such as erythrocytes, leave the cycle at mitosis and remain non-dividing. There are also cells in adult tissues which ordinarily do not divide, but can be stimulated to do so under certain conditions. For instance, liver cells are stimulated to divide after the removal of part of the liver (hepatectomy). However, for most purposes the mammalian cell cycle is studied in cells growing outside the body in tissue culture. For example, lymphocytes* can be studied *in vitro*, where they can be artificially stimulated to divide.

In most plant and animal cells the cell cycle lasts about a day, but only an hour is occupied by mitosis (Fig. 1). The cell cycle of some single-cell organisms is much faster, for example the protozoan[2] *Tetrahymena*. Yeasts can complete the cycle in 2.5–3 hours. The cell cycle of cancer cells is similar to cell cycles of other reproducing cells; the phases of the cycle are the same but the cells repeat the cycle without restraint.

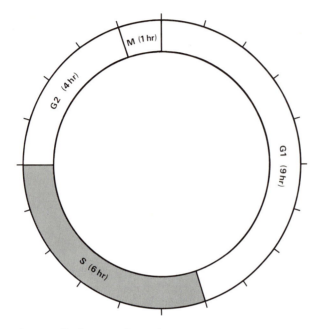

Figure 1 The cell cycle in the mouse L cell showing the duration of the G1, S, G2 and M phases.

In this Unit we will discuss information on the cell cycle obtained from the study of a wide variety of cells including bacteria, lower eucaryotes such as yeasts[2] and slime moulds[3], and mammalian cells. It will become apparent that one of the favourite eucaryotic organisms used to study the cell cycle is yeast. This is because yeasts can be grown as easily and cheaply as the molecular biologists' favourite organism, *Escherichia coli*. Unlike mammalian cells, yeasts can be manipulated genetically with ease. It is as simple to generate and analyse mutants in yeast as in bacteria, and as we shall see later this becomes important in determining which events are indispensable for a cell to proceed through the cell cycle. On the other hand, yeasts have the cellular complexity of higher eucaryotes. They have a nucleus bounded by a nuclear membrane, as well as mitochondria, Golgi bodies and endoplasmic reticulum. In addition, there are several chromosome pairs and they are apportioned to the potential daughter cells according to the classical laws of mitosis.[4]

Mitosis and cell cleavage will be discussed in the TV for Unit 7. In this Unit we will explore interphase and consider the patterns of growth and synthesis during the cell cycle. In the course of a cell cycle a cell must double its cellular components. Once the cell has doubled in size, it divides—provided that certain conditions have been met, for instance, provided that the DNA complement of the cell has doubled. There are no striking morphological changes during interphase other than the enlargement of the cell. This and the fact that RNA and total proteins are synthesized throughout the cycle led many to suppose that interphase was a process of uniform growth with all components increasing at the same rate. However, one of the most interesting advances in our understanding of the cell cycle is that interphase is not a time of uniform growth, but a time when striking changes in the chemical composition of the cell take place.

The most obvious example of this is the doubling in DNA content of the cell over a relatively short part of the cycle. Although total protein increases continuously throughout the cell cycle, the individual proteins that comprise total protein are in many instances synthesized over a very restricted part of the cycle. This is particularly true for a variety of enzymes and is discussed in detail in

---

* Lymphocytes of higher animals contribute to the formation of antibodies.

Section 7.2.4. In addition, there is evidence for the synthesis of histones (Unit 1, Section 1.4.1) only during the S period. There are also examples of periodic synthesis of cytoplasmic proteins; for instance, in human lymphoid cells* and in mouse myeloma** cells, immunoglobulins are synthesized only from late G1 until nearly the end of S. There is then clear evidence of chemical differentiation which precedes the morphological differentiation apparent during mitosis. This will be emphasized in Section 7.2.4.

7.1 Methods for studying the cell cycle

Study Comment After reading Section 7.1 you should appreciate that different principles are involved in induction and selection synchrony. Their advantages and disadvantages should be understood. Remember at least two examples of each method.

A variety of methods has been used to study the mitotic cell cycle of a wide range of cells including bacteria, protozoa, algae, yeasts and higher plant and animal cells. Some studies have been made on individual cells, particularly photomicrographic and cytological studies. However, it is difficult to carry out meaningful biochemical analyses on single cells and this has necessitated the development of methods to obtain synchronous cultures, in which all cells are at the same stage of the cell cycle and all cells in the population divide at the same time. In these circumstances the events taking place in single cells are amplified and made amenable to biochemical analysis.

There are two ways to obtain synchronous cultures; each method has advantages and disadvantages.

7.1.1 Induction synchrony

Cells at different stages of the cell cycle can, by a variety of treatments, be blocked at a particular stage of the cell cycle. One instance of this is to inhibit DNA synthesis in populations of cells until all cells are in a pre-DNA synthetic period. The DNA block is then released and synchronous growth follows (see Fig. 2). An

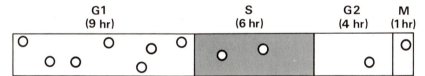

In an exponential culture, there are cells at all stages of the cell cycle. N.B. There are twice as many cells early in the cell cycle as late.

Addition of a DNA synthesis block results in cells piling up at the end of G1.

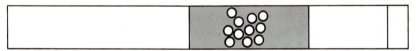

If the block is removed, the cells progress through S phase and the rest of the cycle synchronously.

As time progresses, the cells begin to lose their synchrony.

Figure 2 Induction synchrony by DNA blocking agents. After release from the blocking agent and subsequent cell division, the number of cells doubles.

* LYMPHOID CELLS Cells derived from lymphoid tissue (e.g. thymus) which are found throughout the lymphatic and blood systems of the body.

** MYELOMA CELLS Cells from a plasma cell tumour, often invading the bone marrow.

7

alternative procedure, which is discussed at greater length in Section 7.5.1, uses repeated heat shocks to obtain synchrony from asynchronous cells. Here cells are subjected to a series of temperature increases at intervals that correspond to the length of the cells' cycle—synchronous growth ensues. Methods in which cells are forced to attain a common stage of the cell cycle have become known as 'induction synchrony' methods or 'phasing' methods. Some examples are shown in Table 1 (a) on p. 10 which summarizes many of the techniques used.

induction synchrony methods

7.1.2 Selection synchrony

An alternative way to obtain synchronous cultures is to select cells at a particular phase of the cycle from populations containing cells at different stages. Such methods exploit the fact that the physical properties of cells change during the mitotic cell cycle. In the fission yeast*, a simple eucaryote, the cell increases in size throughout the cell cycle (see Fig. 3). Therefore, if cells of a particular size

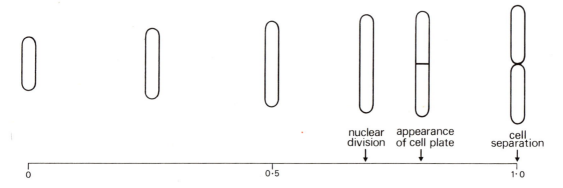

nuclear division — appearance of cell plate — cell separation

0   0·5   1·0

stage of cell cycle

can be separated from a culture growing exponentially they should be at the same stage of the cell cycle and will divide synchronously if inoculated into fresh medium. Mitchison and Vincent (1965) developed a method to accomplish this. A 1-litre culture of yeast cells is grown overnight and harvested by filtering the cells from the growth medium and concentrating them in about 5 cm³ of fresh growth medium. Cells are then layered on to a linear sucrose gradient in a centrifuge tube (see Fig. 4) using a pipette. The tube is centrifuged until the largest cells have moved approximately two-thirds of the way down the tube (Fig. 4). The uppermost layer of cells, corresponding to small cells early in the cell cycle, is removed. These cells are then inoculated into fresh culture medium and synchronous divisions occur. This basic method has also been used with bacteria, algae and mammalian cells and can be used for almost any free-living cell which increases in size during the mitotic cell cycle. An alternative procedure for mammalian cells in tissue culture depends on the fact that cells in the metaphase[5] stage of the mitotic cycle can be detached from glass in monolayer cultures. Thus, if tissue cultures are grown as monolayers of cells on glass and the flasks are shaken gently, only cells in metaphase detach from the glass; these can be removed and used to initiate a synchronous culture.

Figure 3 The cell cycle of *Schizosaccharomyces pombe*, the fission yeast. This organism increases in length during the cycle.

A particularly ingenious method of selecting synchronous cultures of the unicellular protozoan *Tetrahymena* has exploited the observation that these organisms are indiscriminate feeders and feed for all but a very short stage of the cell cycle. Thus, small particles of iron are fed to a culture for a short time and the cells are then separated according to density. Those cells that had engulfed the particles are denser than the rest of the population and can be removed. The less dense cells which had not engulfed the iron particles can be used to inoculate a synchronous culture.

Methods involving the separation of cells at a particular stage of the cell cycle from a growing culture have been termed 'selection' methods. Some examples are shown in Table 1 (b) on p. 10.

selection methods

* In contrast to budding yeasts, fission yeasts reproduce by dividing in half.

It is of vital importance that any culture employed to study the cell cycle should exhibit balanced growth. This was emphasized in 1957 by Campbell, who defined growth as 'balanced over a time interval if, during that interval, every extensive property of the growing system increases by the same factor'. Biochemical events should repeat in successive cell cycles at the same stage in each cycle. Balanced growth may not be achieved in synchronous cultures if the environment of a cell is changed drastically by the conditions used to obtain synchronous growth. Phasing methods all rely on changes in the environment, such as temperature, illumination or nutrition, to cause synchronization. It is perhaps inevitable that they often result in metabolic disturbance and perhaps an abnormal cell cycle. For this reason such methods are less commonly used nowadays, but sometimes they are the only methods available. They do have the advantage that large numbers of cells can be made synchronous. As selection methods depend on the existence for a short time in the cell cycle of a specific physical characteristic which allows cells at a particular stage of the cycle to be isolated, it follows that the yield from a culture that contains cells at all stages of the cycle will be small. Phasing methods, on the other hand, synchronize all the cells in a population.

balanced growth

Recently, zonal centrifugation (Unit 1, Section 1.2) of exponentially growing cultures has been used in some instances to replace synchronous methods. The method has been applied to cells, such as bacteria, yeasts and mammalian cells, that grow in size throughout the cell cycle. The method in principle is very similar to that shown in Figure 4. It is based on the fact that, during zonal

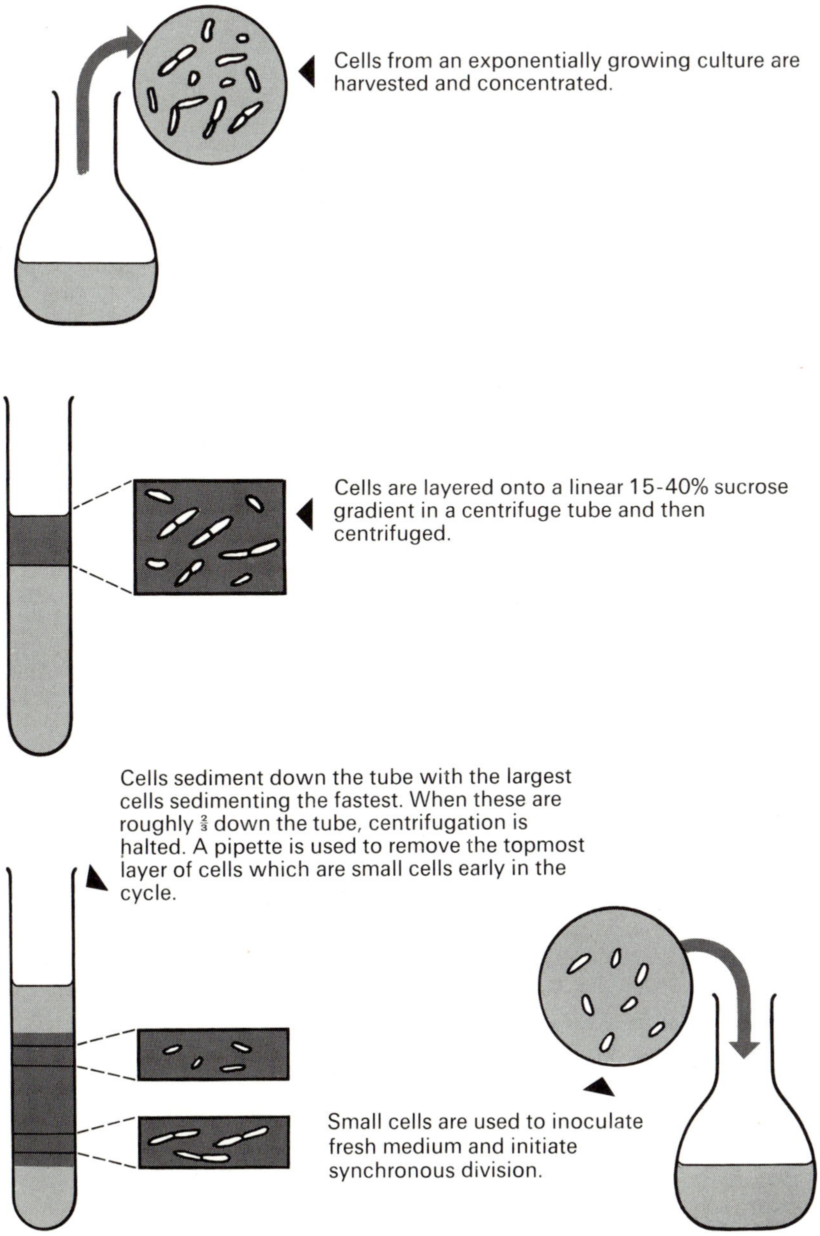

Cells from an exponentially growing culture are harvested and concentrated.

Cells are layered onto a linear 15-40% sucrose gradient in a centrifuge tube and then centrifuged.

Cells sediment down the tube with the largest cells sedimenting the fastest. When these are roughly ⅔ down the tube, centrifugation is halted. A pipette is used to remove the topmost layer of cells which are small cells early in the cycle.

Small cells are used to inoculate fresh medium and initiate synchronous division.

Figure 4 The establishment of synchronous cultures of the fission yeast, using selection synchrony.

centrifugation in a gradient, cells tend to separate according to their size. As they increase in size throughout the cycle, cells are also separated according to their stage in the mitotic cell cycle. Therefore, fractions taken across the zonal rotor will contain cells early in the cell cycle near the centre of the rotor and increasingly older cells farther towards the edge. The complete cell cycle can be analysed in this way. Analysis of exponential cultures after separation on the zonal rotor avoids many of the difficulties associated with synchronous cultures. There can be no artefacts caused by metabolic disturbance as in phasing methods, and one can analyse all of an exponential culture, not just a fraction of it as in selection synchrony methods.

There are some instances of natural synchrony within populations. The best-known example in the animal kingdom is the early embryo, although it is unusual in that many divisions occur without concomitant growth. The number of divisions after fertilization that remain synchronous varies from three in the sea urchin, to four in the mouse, to nine in the holothurian *Synapta digitata*.

Table 1 Methods of obtaining synchronous cultures

(a) Induction synchrony (or phasing methods)

Method	Organisms	Principle
Light cycles	Algae such as *Euglena*, *Chlorella*, *Chlamydomonas*	Cultures of photosynthetic organisms are treated with alternate periods of light and dark. If these periods are coincident with the doubling time, division occurs once a cycle.
Nutritional methods	Bacteria and yeasts, mammalian cells, higher plant cells	Cells are grown until nutrients are exhausted and re-inoculated in fresh medium. Sometimes growth on minimal medium with a shift to a richer medium results in synchronous division. It is assumed that starved cells arrest at the same stage of the cell cycle.
Temperature changes	Algae, yeasts, bacteria, *Tetrahymena*	These involve a temporary shift from an optimal temperature to a high temperature. Sometimes a single shift is adequate for synchrony, but usually repeated shifts are necessary. (See Section 7.5.1.)
Inhibitor blocks	Mammalian cells, yeasts, *Tetrahymena*	An inhibitor acting usually on either DNA synthesis or mitosis is added to growing asynchronous cultures. Cells accumulate at one stage of the cell cycle and after release from the inhibitor divide synchronously (Fig. 2).

(b) Selection synchrony

Method	Organisms	Principle
'Wash-off'	Mammalian cells, such as chick embryo cells, HeLa cells, Chinese hamster ovary cells.	Tissue culture cells growing as a layer on glass loosen the attachment during the metaphase of mitosis and can be separated by gentle washing-off.
Sedimentation size	Mammalian cells, yeasts, bacteria, *Chlorella*	This method depends on cells being separated according to size upon centrifugation. Small cells of the top layer are removed to start a synchronous culture.
Sedimentation density	Yeasts, *Chlorella*	Cells with different densities at different stages of the cell cycle can be separated by centrifugation to equilibrium in a density-gradient.
Adsorption and 'grow-off'	*Escherichia coli* B	Cells are adsorbed on to a filter. When the cells divide, the daughter cell detaches and can be used to initiate sychronous cultures.

7.1.3 The measurement of cell synchrony

The development of synchronous cultures is usually monitored by following the doubling of cell numbers. This can be done electronically using a counter of the type originally designed to count blood cell numbers. A typical synchronous culture is shown in Figure 5. As shown, in practice there is never 100 per cent synchrony, and so cell numbers do not rise like steps in a staircase. In addition, synchrony decays with time so that after three or four generations, little or no synchrony exists and cell numbers rise exponentially, rather than in the step-by-step manner associated with synchrony. Less than perfect synchrony and its decay are a result of variability in the generation times of individual cells in the population.

In a number of organisms synchronous cultures cannot be monitored by measuring cell numbers; this is particularly true of cells that do not separate after

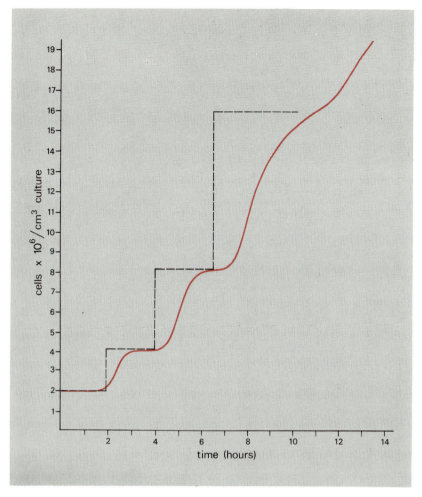

Figure 5 A synchronous culture of *Schizosaccharomyces pombe*, a fission yeast, prepared by separating by centrifugation an exponential culture on the basis of size and using small cells from the centrifuge tube to start the synchronous culture. The dashed line shows the theoretical values for 100 per cent synchrony where every cell is at exactly the same point in the cell cycle: hence the doubling of numbers occurs with a vertical slope. The red line shows the actual increase in the number of cells.

mitotic division. In the acellular slime mould *Physarum polycephalum** for instance, there are successive, very synchronous, mitotic divisions without cell separation so that a multinucleate plasmodium forms. In such cases, the cell cycle is defined by cytological markers.**

When you have completed Section 7.1, attempt SAQs 1–4 on p. 31.

Summary of Section 7.1

1   The cell cycle is usually studied using synchronous cultures.

2   There are two basic methods:
(a) *Induction synchrony*, in which all cells in the population are forced to attain a common stage in the cell cycle;
(b) *Selection synchrony*, in which cells at a particular stage in the cell cycle are selected from populations containing cells at different stages of the cell cycle. This method exploits the fact that the physical properties of cells change during the cell cycle.

3   Induction methods give a greater yield of cells than selection methods, but the cells may have abnormal properties because of metabolic disturbance created by the synchronization procedure.

* Those who have done S2–5, do not confuse *Physarum* with the *cellular* slime moulds such as *Dictyostelium*. *Physarum* forms an aggregate where all the cells fuse to form a single multinucleate (syncitial) body. It is a member of the so-called 'true' slime moulds, which are a group distinct from the cellular slime moulds in which the aggregate and later stages are made up of many individual cells.

** MARKER Mitchison defines cycle 'markers' as discrete events, chemical, structural or physiological, which happen at a particular point in the normal cycle. Some examples are the beginning and end of the S period and the stages of mitosis. Other examples are given in Section 7.5.3.

7.2   Cell growth

Study Comment Section 7.2 deals with the growth of cell components and organelles during the cell cycle. You should understand why patterns of synthesis of cell components can be more readily detected by studying rates of synthesis rather than total quantity present at any one time. Remember the rationale behind experiments to determine whether organelles divide or form *de novo*.

Section 7.2.3 deals with the regulation of the periodic synthesis of enzymes and two models which have been proposed to explain it. The linear transcription model is simple in concept, but some of the experiments designed to test it are subtle and depend upon a genetical knowledge of the organism.

7.2.1   Criteria for measuring cell growth

In general, cells grow and increase in mass during the cell cycle. There are, however, some instances of cell division without any growth with the result that the daughter cells are relatively small. This occurs during the cleavage process of a developing embryo.

We can see from Figure 6 that there are a number of different criteria which can be used to measure cell growth. A rigorous study of cell growth would include data on many of these criteria as measurements of any one might be misleading. For instance, an increase in volume might be primarily caused by either macromolecular synthesis or water uptake.

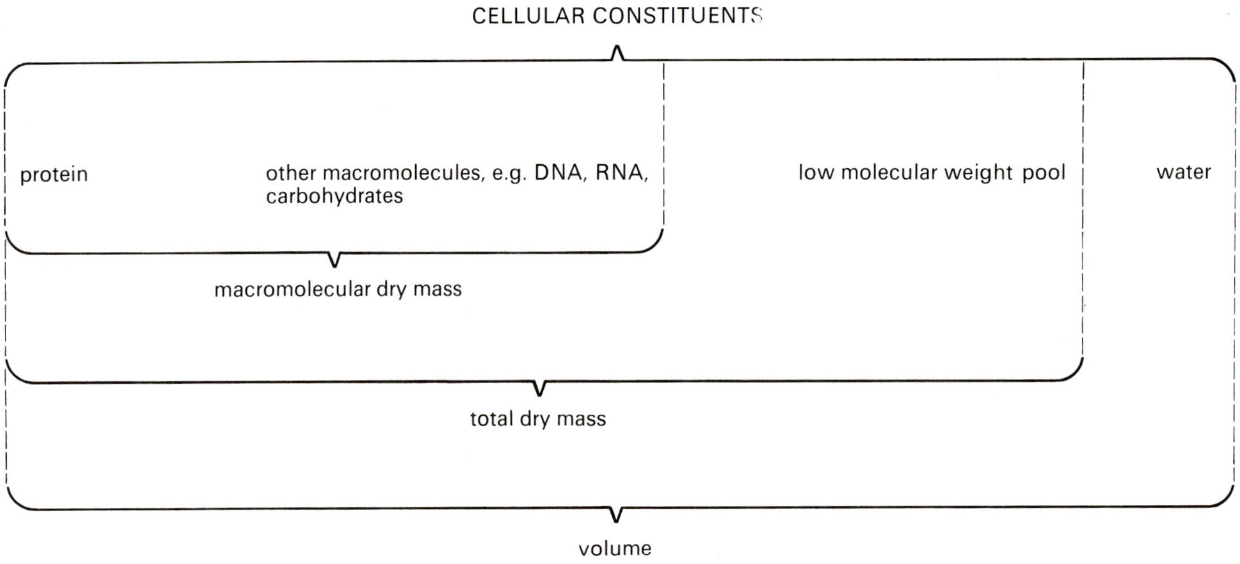

CELLULAR CONSTITUENTS

Figure 6 Criteria for measuring cell growth can involve monitoring increases in these quantities.

In most cells studied, volume, dry mass and total protein exhibit continuous increases throughout the cell cycle. It is, however, difficult to find the exact pattern of synthesis of a cell component from measurements of the total quantity of that component throughout the cycle. This is illustrated in Figure 7 which shows the increase in quantity during the cell cycle of a component which increases (a) linearly, (b) exponentially and (c) linearly with a rate doubling in the middle of the cell cycle. Given some scatter in experimental measurements it can, in practice, be very hard to distinguish between these three possibilities.

However, the problem is easier when the investigator uses an isotopically labelled precursor which is incorporated into the macromolecule being studied. For instance, cells at different stages of the cell cycle can be exposed to ^{14}C-labelled arginine for a small fraction of the cell cycle and the amount of label found in protein in different samples can be determined. Thus the *rates* of protein synthesis at different stages of a cell cycle are measured. Figure 8 shows that different patterns of synthesis can more readily be distinguished by studying rates than by studying quantity (Fig. 7).

Protein synthesis in mammalian cells has been followed in this way and in a variety of such cells it has been observed that protein synthesis is continuous and that its rate increases throughout the cycle. Similarly, RNA synthesis can be studied by labelling cells with radioactive uridine, a precursor of RNA, for a short fraction of the cycle at different stages. Such experiments demonstrate that in the higher eucaryotes RNA synthesis occurs at all times during the cell cycle, except during the metaphase stage of mitosis when the DNA in the chromosomes is in a highly condensed state and presumably cannot be transcribed. The cessation of RNA synthesis is not observed in yeasts which are lower eucaryotes with a primitive mitosis. Relatively little work has been done on the synthetic pattern of the various classes of RNA (transfer, messenger and ribosomal[6]).

Earlier it was mentioned that a variety of growth parameters should be measured for an understanding of cell growth. This is emphasized by the finding that not all of these parameters increase in the same manner. For instance, it has been shown that in yeasts, dry mass increases linearly throughout the cell cycle. On the other hand, cell volume increases in an S-shaped (sigmoid) curve, almost forming a plateau at the beginning and end of the cell cycle. Because dry mass and volume increase in a different pattern, cell density must change during the cell cycle with a maximum early in the cell cycle and a minimum later in the cycle. (This fact has been exploited in obtaining synchronous cultures of yeast. Exponential cultures are harvested and centrifuged in a density gradient to equilibrium (Unit 1, Section 1.2). Cells from the top layer (lowest density) or from the bottom layer (densest cells) can be used to initiate a synchronous culture.)

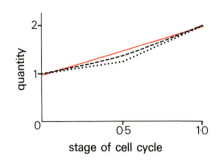

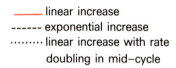

—— linear increase

------ exponential increase

·········· linear increase with rate doubling in mid–cycle

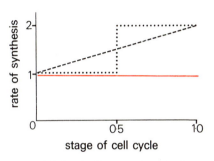

Figure 7 (*above*) Pattern of synthesis expressed as a change in quantity of a macromolecule during the cell cycle.

Figure 8 (*below*) Pattern of synthesis expressed as a rate of synthesis of a macromolecule during the cell cycle.

7.2.2 Growth of cell organelles

Having considered some of the patterns of overall growth during the cell cycle, let us consider the growth of some cell organelles. Clearly the nucleus must replicate once during a cell cycle. It appears that in contrast with overall cell growth, the nucleus increases in dry mass and volume in the second half of the cell cycle, before cell division. In higher cells the nuclear membrane breaks down during prophase and is reconstituted in telophase.[7] In some lower eucaryotes such as yeast, the nuclear membrane does not break down during mitosis.

What of the cytoplasmic organelles, such as the mitochondria, plastids and vacuoles? What is the pattern of their growth during the cell cycle and are they formed by the division of pre-existing organelles rather than by growth of rudimentary precursors? If organelles replicate, how is their division integrated into the mitotic cycle?

There is microscopic evidence in a variety of lower cells that mitochondria appear to divide and, in some Protozoa, there is but one mitochondrion which replicates at cell division. Microscopic evidence is, however, difficult to evaluate because mitochondria are not perfectly round and therefore thin sections through parts of a dumb-bell-shaped mitochondrion would give the appearance of two mitochondria.

Luck (1963) used a short pulse of radioactive choline to investigate whether mitochondria divide or form from rudiments. The label is incorporated into the phospholipid of mitochondrial membranes. After a short exposure to radioactive choline, cells were grown in the presence of unlabelled choline. Samples were taken at intervals and mitochondrial preparations were analysed for the incorporation of the radioactive label.

QUESTION What conclusions would Luck have made:
(a) if, with time, a decreasing proportion of the mitochondria retained the original label and an increasing number were unlabelled?
(b) if the label were randomly distributed throughout the mitochondrial populations with steadily decreasing label as time elapsed?

ANSWER If Luck had observed (a) he would have concluded that mitochondria form *de novo*. However, he observed (b) and concluded that mitochondria divide, and newly divided mitochondria grow and divide again, diluting out the labelled choline from the whole mitochondrial population.

There is microscopic evidence in lower plants that chloroplasts also divide. However, in higher plants, chloroplasts develop from proplastids and do not divide in differentiated cells, but differentiated plant cells don't divide either.

How is the division of cell organelles integrated with the cell cycle? Do mitochondria all divide at one time (synchronously) during the cell cycle? This is an area that is beginning to receive some attention and there is evidence for synchronous mitochondrial division in the growing hyphae of *Neurospora* as well as during synchronous growth of the fission yeast *Schizosaccharomyces pombe*. Centrioles (Unit 1, Section 1.4.7) are unlike mitochondria and chloroplasts in that they arise *de novo*. A new centriole forms adjacent to the mature centriole during early S phase in sea-urchin embryos or mammalian cells.

7.2.3 Growth of cell surfaces

Unlike animal cells, plant cells possess a well-defined cell wall. A major technique in the study of its growth is to label cells with cell-wall precursors for a brief period and then locate the growing regions that contain label by autoradiography. Such techniques have revealed that in many higher plant cells, growth takes place throughout the cell wall and is not localized. There are, however, exceptions to the rule, notably in the lower plants, fungal hyphae and yeasts, where tip growth is observed. The addition of new wall at the tip of *Schizosaccharomyces pombe*, a fission yeast, increases exponentially with time and in doing so, mirrors the increase in cell volume.

Growth of the cell wall of bacteria has been studied by immunofluorescence (see Unit 1, Section 1.1.1). Antibodies are prepared against cell wall material and the antibodies themselves are labelled with a fluorescent dye. The fluorescent antibody is then applied to growing cells for a fraction of the cell cycle; after this, the antibody is removed and the cells are observed using a fluorescence microscope. The cell wall made at the time of the addition of the antibody will fluoresce, whereas new cell wall made after the removal of the antibody will not. It is best to confirm this using the converse conditions. Cells are first exposed to unlabelled antibody, then grown for a period and finally exposed to labelled antibody. In this case, old wall will be dark and only new wall will fluoresce. Using such techniques it has been observed that in *Streptococcus pyogenes* new cell wall is formed equatorially and old material is conserved. The implications of this are discussed in Section 7.3.2.

Although its significance is not yet appreciated, the cell surface of Chinese hamster ovary cells changes during the cycle. Cells are spherical during mitosis but, as the cells enter G1, they flatten and their cell surface becomes covered with blobs and finger-like microvilli.[8] Further into G1 the surface is covered in ruffles.[9] During the S phase, the surface flattens and becomes smooth, but ruffles form again during G2. It has been possible to exploit the changes in physical properties of the cell surface to make synchronous cultures of these and other mammalian cells. During mitosis, when cells become spherical, they can easily be detached from the surface on which they are growing. If a culture growing on a surface is washed, only the cells in mitosis are flushed out and these can be used to initiate a synchronous culture (see Table 1 on p. 10).

7.2.4 Synthesis of individual proteins

We have seen from Section 7.2.1 that, in terms of size, protein and RNA content, the cell appears to undergo continuous uniform growth during interphase. The cell does not appear to be changing the relative levels of its various components. However, we mentioned in the Introduction that total cell protein is composed of thousands of individual proteins, and evidence will be presented in this Section that by no means all of these proteins are synthesized continuously. The periodic synthesis of histones and immunoglobulins was mentioned in the Introduction. Now we shall survey the patterns of synthesis of individual enzymes during the cell cycle.

Figure 9 shows some of the patterns of individual enzyme synthesis that have been observed in studies of the cell cycles of bacteria and a number of eucaryotes. The synthesis of most enzymes in eucaryotes resembles Figure 9(a). This pattern is observed for stable enzymes that are synthesized periodically once per cell cycle, so that enzyme accumulates in the cell in a step-like manner. In various budding yeasts for instance, over 30 enzymes have been observed to accumulate in this way. There is for each individual enzyme a characteristic stage during the cell cycle when it is synthesized, and the enzyme steps are repeated in successive cell cycles at the same stage. Steps in enzyme synthesis are not restricted to a particular part of the cycle, and for different enzymes, steps may occur at different stages of the cycle. Periodic synthesis of unstable enzymes results in the peak pattern in Figure 9(b). Enzyme activity increases during the period of enzyme synthesis and declines when no enzyme is synthesized. Some enzymes are synthesized continuously throughout the cell cycle, in such a way that enzyme accumulates as in Figure 9(c) or 9(d). It is experimentally very difficult to decide if enzyme accumulation is exponential (Fig. 9(c)) or linear with a constant rate of synthesis until a particular stage in the cell cycle when the rate doubles (Fig. 9(d); see also Fig. 7 on p. 13).

We know that individual enzymes are synthesized periodically; however, total protein increases continuously throughout the cell cycle. It seems that individual proteins must be synthesized at different times of the cell cycle, but some protein synthesis takes place throughout the cycle to give a continuous increase of total protein.

In bacteria and lower eucaryotes (such as yeasts and slime moulds) as well as in mammalian cells, about two-thirds of those enzymes studied show periodic synthesis. The regulation of this periodic synthesis of enzymes has been the subject of much study and two theories have been proposed to explain it:

(a)   The oscillatory repression model

This theory is supported mainly by work on bacteria and derives from regulation by feed-back control and the control of enzyme synthesis by inducers and repressors.[10] One of the consequences of feed-back repression is that enzyme synthesis can oscillate. Enzyme synthesis leads to an increased rate of formation of end-product. When this end-product reaches a critical concentration, enzyme synthesis is repressed. During subsequent growth and metabolism, the concentration of end-product decreases until it falls below a critical concentration, at which point enzyme synthesis is renewed (see Fig. 10).

(a)

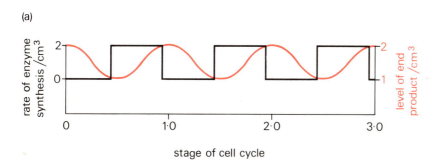

(b)

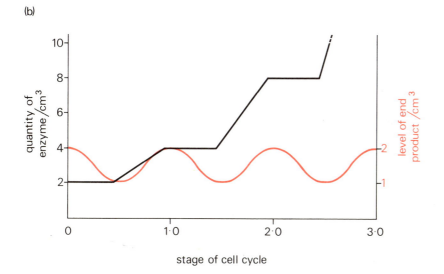

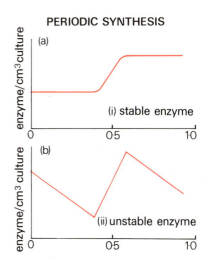

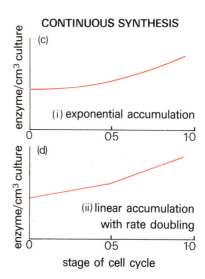

Figure 9 (*above*) Variations in the pattern of individual enzyme synthesis.

Figure 10 The level of end-product oscillates with a minimum halfway through the cycle and a maximum at the beginning of the cycle. Enzyme synthesis is repressed at 1.0, 2.0 and 3.0, when the end-product reaches a critical concentration. Enzyme synthesis is turned on again at 0.5, 1.5 and 2.5, when the end-product declines to a critically low concentration. Renewed enzyme synthesis results in the concentration of end-product rising again until the critical concentration for enzyme repression is reached.

Periodic synthesis will occur when the enzyme is partially repressed or induced. If the enzyme is fully induced or completely repressed at basal level, periodic synthesis should not be observed and the enzyme will be synthesized continuously. The theory fits most of the data for bacterial studies of enzyme synthesis and the model allows a number of predictions, which have been tested. The removal of repressor or the addition of inducer should lead to enzyme synthesis if feed-back repression or induction is controlling synthesis. Thus, in synchronous culture, the addition of inducer at any time during the cell cycle should result in enzyme induction. Further, the rate induced as a result of induction should double when the gene for that enzyme duplicates as a consequence of DNA replication. This prediction is upheld: if thiomethylgalactoside (a chemical analogue of lactose and an inducer of β-galactosidase) is added to synchronous cultures of *Escherichia coli* at different stages of the cell cycle, β-galactosidase enzyme is induced. The rate of induced synthesis doubles sharply at the point where the structural gene for that enzyme doubles. If nalidixic acid (an inhibitor of DNA synthesis) is added to a synchronous culture of *E. coli* then enzyme is induced (upon the addition of inducer), but in the absence of DNA synthesis no doubling in the rate of production of induced enzyme occurs. This suggests a method of mapping the order of genes on a chromosome (provided replication of DNA follows gene order). If we consider several induced enzymes and plot the points at which their rate of synthesis doubles, this should accurately reflect the order of the respective genes on the chromosome and the spacing between them. So, independent genetic techniques of mapping the order and spacing of genes on chromosomes can be used to check the oscillatory repression model—the model is known to hold true for six enzymes in *E. coli*.

Although there is an impressive amount of evidence in favour of the oscillatory repression model in bacteria, it is very difficult to formulate a definitive test. For example, if measurements of the end-product present in the cell (end-product pool) were made and found not to fluctuate, then it could still be argued that the pool may be only a small fraction of the total and therefore oscillations in it will go unobserved. Alternatively, if the end-product pool is found to fluctuate it does not prove that these oscillations result in periodic enzyme synthesis. Indeed the converse might be true.

(b) The linear transcription model

An alternative theory accounting for the periodic regulation of enzyme synthesis has been advanced to explain data obtained from cell-cycle studies on budding yeasts. The development of the theory arose from some experimental data observed in yeast which are incompatible with the oscillatory repression model. In particular, experiments have been done with a mutant of *Saccharomyces cerevisiae* that produces UDP galactose epimerase constitutively, that is, the synthesis of this enzyme, a member of the galactose utilization pathway, is not subject to the influence of inducers or repressors. This being so, there is no reason to expect periodic enzyme synthesis because oscillations in the level of effectors will have no influence on enzyme synthesis. However, it has been observed that the synthesis of UDP galactose epimerase is periodic and occurs at exactly the same stage of the cell cycle as the wild type. Therefore, at least in yeast, there must be some different control system regulating periodic enzyme synthesis.

It is not perhaps surprising that eucaryotic organisms should have evolved regulatory mechanisms additional or alternative to those used by procaryotes. The eucaryotic cell is generally larger than the procaryotic cell; it is divided into compartments, for example, nuclear and mitochondrial, and has larger pools of macromolecular precursors. Thus the eucaryotic cell is more effectively buffered from changes in the external environment and its immediate response to such changes is less vital. A biochemical illustration of these differences in response to environmental changes is that whereas in bacteria, enzyme levels sometimes change one-thousandfold in reponse to changes in the external environment, changes in the enzyme levels of eucaryotic cells are generally very much less—a factor of about $\times 10$.*

---

* Having said that there may well be differences in metabolic regulation between eucaryotic and procaryotic organisms; there may also be differences among eucaryotes. In particular, there may be differences between cells that grow unicellularly, for example, between protozoa, yeasts and some algae and those cells that exist in organized tissues.

We have mentioned previously that there is a temporal synthesis of individual enzymes during the cell cycle: that enzymes are synthesized once per cycle at a particular stage of the cell cycle which is characteristic for each enzyme. From his work on yeast, H. O. Halvorson has suggested that the simplest model to provide a programmed expression of the genome is the chromosome itself. Simply, this model (the linear transcription, or linear reading, model) suggests that the order of genes on a chromosome determines the programme for transcription and translation during the mitotic cell cycle. There is linear transcription of a part or the whole of a chromosome. Thus there is a linear relationship between the time of enzyme synthesis during the cell cycle and the position of the structural genes of those enzymes along the chromosome.

Using this model various predictions can be made and tested. For instance, there should be one period of enzyme synthesis per structural gene per chromosome. In diploid strains of yeast only one period of enzyme synthesis per cell cycle is observed for either homozygous (M_1M_1) or heterozygous (M_1m_1) strains of yeast indicating that gene dosage at one locus does not effect the timing of enzyme synthesis. However, if strains are constructed containing additional non-allelic genes (M_2 and M_3) (i.e. additional genes at different positions on the chromosomes), then the number of periods of enzyme synthesis corresponds to the number of non-allelic M genes present. This suggests that non-allelic genes may be transcribed at different points of the cell cycle. Another prediction of the linear transcription model is that the interval between the expression of genes on the same chromosome should be related to their distance apart on the chromosome. In another strain of yeasts, *Saccharomyces lactis*, β-galactosidase and β-glucosidase, two enzymes concerned with sugar breakdown, are very closely positioned on the chromosome. This is apparent from the very small recombination[11] observed for these genes (1 per cent). (The farther apart two genes are on the chromosome, the greater the chance of crossing-over and recombination.) Thus one would expect the periods of synthesis of these two enzymes to be almost identical. In fact they are both synthesized periodically at almost the same point in the cell cycle.

A knowledge of yeast genetics has made possible a study of four enzymes whose genes are located on chromosomes V of yeast. Each of the four enzymes is synthesized periodically and in the same order that the structural genes for the enzymes are situated on the chromosome (see Fig. 11).

gene order on chromosome Ⅴ

time during the cell cycle of periodic enzyme synthesis

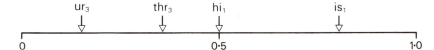

Figure 11 Linear transcription.

More convincing evidence might be observed if the timing of enzyme synthesis could be studied in strains in which the positions of genes on chromosomes were altered either by translocation to another chromosome or by inversion of part of a chromosome.* Cox and Gilbert (1970) have studied a strain of yeast which has a double inversion in chromosome II. Here a segment of the chromosome containing genes for acridine resistance (ac_1) and galactokinase (gal_1) is inverted, together with another part of the chromosome containing α-amino adipic acid reductase (lys_2—an enzyme in the lysine biosynthetic pathway) and genes of the tyrosine (tyr_1) and histidine (his_7) biosynthetic pathway. This results in the genes for galactokinase (gal_1) and α-amino adipic acid reductase (lys_2) being much further apart on the chromosome than in the wild type

---

* A translocation involves the detachment of part of a chromosome and its attachment to another chromosome. An inversion occurs when a segment of a chromosome detaches and rejoins the rest of the chromosome but the other way round so that the order of genes on the inverted segment is reversed in relation to other parts of the chromosome.

(see Fig. 12). When the synthesis of these enzymes was examined in synchronous cultures it was observed that both enzymes were synthesized periodically in the inverted strain as well as the wild-type strain, but the difference between the steps had increased from 0.15 of a cell cycle in the wild type to 0.5 of a cell cycle (Fig. 12).

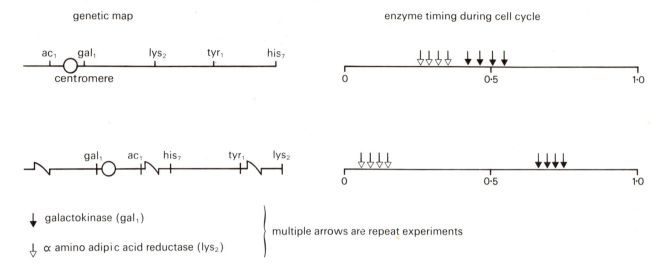

In summary, periodic enzyme synthesis is a feature of both the procaryotic and the eucaryotic cell cycle. The oscillatory repression model fits much of the data on enzyme synthesis in bacteria but is inadequate as an explanation of the phenomenon in yeasts. Although the linear transcription model has a good deal of supporting evidence in yeast, an additional mechanism must be postulated to account for immediate enzyme synthesis upon induction at any stage of the cell cycle. In addition, there is evidence for translational controls of enzyme synthesis in eucaryotes, particularly in developing embryos.[12]

Figure 12 The effect of a double inversion in chromosome II on enzyme timing.

It is not surprising that the regulation of enzyme synthesis during the cell cycle of eucaryotes is complex. In one of the simplest systems studied, the regulation of bacteriophage λ, at least 15 positive and negative controls govern transcription and it would be naïve to expect a eucaryotic cell to be any less complex. Our lack of understanding of temporal control of enzyme synthesis during the cell cycle is a reflection of our poor understanding of transcription and translation in any eucaryotic cell. Our studies would be enhanced by some knowledge of the half-lives of individual messenger RNAs and of the time during the cell cycle at which *de novo* synthesis of inactive protein and its conversion to active enzyme takes place.

The two theories discussed so far are of interest because they relate to the question of why parts of the genome are expressed at one point in time and not another. This crucial point relates both to the cell cycle and to differentiation where parts of chromosomes remain unexpressed during certain developmental stages but result in protein synthesis at others.

Finally, it is interesting to speculate why periodic enzyme synthesis is so widespread. You might think that continuous synthesis is a better way to increase the cell components during the mitotic cycle without upsetting the balanced growth of the cell. What are the implications of periodic enzyme synthesis for cell metabolism? We can only speculate. A doubling in the level of enzyme over a short period during the cell cycle might have a profound effect on the cell's metabolism. However, if these enzymes are always present in the cell in excess, then no metabolic disturbance will result from a doubling in enzyme level.

When you have completed Section 7.2, attempt SAQs 5–9 on pp. 31 and 32.

Summary of Section 7.2

1 There is a variety of criteria for measuring cell growth.

2 Data on many of these criteria are vital to fully understand growth as any one approach might be misleading.

3   Mitochondria divide during the cell cycle rather than form *de novo*, but the integration of the division of these and other organelles into the cell cycle is poorly understood.

4   Growth of cell wall can be observed by pulse labelling using isotopes or immunofluorescent dyes.

5   Although some enzymes are synthesized throughout the cell cycle the majority are synthesized periodically. There is for each enzyme a characteristic stage during the cycle where it is synthesized and the enzyme steps are repeated in successive cell cycles.

6   Two models—the 'oscillatory repression model' and the 'linear transcription model'—have been proposed to explain periodic enzyme synthesis.

7   We have presented evidence in favour of these theories. They are important in a wider context than the cell cycle because they relate to the question of why parts of the genome are expressed at one point in time and not another. This question is vital to an understanding of the cell cycle and cell development.

7.3   Control of DNA synthesis and cell division in bacteria

> **Study Comment**   This Section discusses the way in which DNA replication is integrated with the bacterial cell cycle. The model of Helmstetter and Cooper is not easy to understand; you will probably need to read through the explanation several times. The key to the model is that in fast-growing cells new rounds of DNA begin to be replicated before previous rounds are complete.
>
> *Study Figure 13 and its caption carefully and relate it to the in-text explanation of the model.*

7.3.1   DNA replication

Most of what we know about the physiology of bacteria has been gathered by studying one organism, *Escherichia coli*. This is particularly true of work on the bacterial cell cycle.

Escherichia coli has a single chromosome. It is circular and about 1 200 μm in circumference: the *E. coli* cell is itself only 1 μm in length! Many aspects of the mechanism of DNA replication remain unknown but its integration with the bacterial cell cycle has been well investigated.

DNA replication always begins at the same point on the chromosome and proceeds in both directions until replication of the whole chromosome is complete. The process is shown diagrammatically in Figure 13. At a variety of growth rates, replication takes 40 minutes. However, the generation time of the cells (the time it takes for cell mass to double) may vary from 20 to 60 minutes. DNA replication may, in cells with a 20-minute generation time, take two cell cycles to complete. How then is it possible for growing cells to divide every 20 minutes and yet for each daughter cell to receive a complete genome? In cells with doubling times of less than 40 minutes new rounds of replication begin before old rounds are complete. Thus, in cells with a generation time of 20 minutes a new round of replication is initiated at the origin every 20 minutes even though the previous replication round is incomplete (see Fig. 13(c)).

Cooper and Helmstetter have studied the timing of initiation of DNA synthesis in cells growing with generation times between 20 and 60 minutes. As the growth rate changed so did the time during the cell cycle when initiation of DNA synthesis took place. A summary of their results and the model they postulate is shown in Figure 13.

The model (Fig. 13) proposes that in cells with doubling times of less than 60 minutes, chromosome duplication (C in their terminology) takes 40 minutes (i.e. C = 40 minutes). Also, there is an additional constant D of 20 minutes which is the time between the end of a round of replication and the following cell division. This can be seen most clearly in Figure 13(a) showing cells with a 60-minute doubling time. Here, chromosome duplication is completed in 40

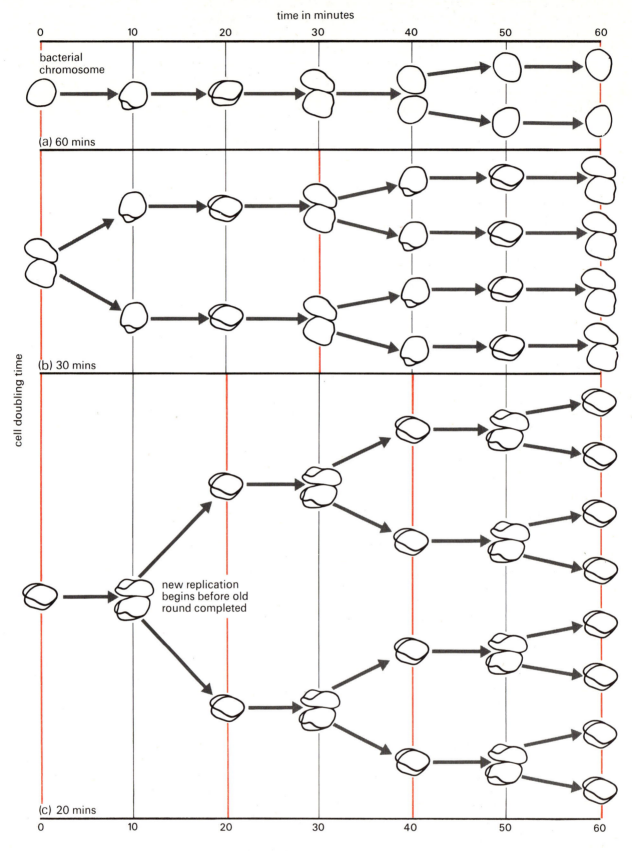

Figure 13 Pattern of chromosome replication over a 60-minute period in cells cultured at different growth rates. The double helix is represented as a single line and the bacterial chromosome as a circle. In every case, each pair of replication forks takes 40 minutes to travel around the chromosome from the origin to the terminus. A new round of replication is initiated at intervals equal to the mass doubling time (i.e. the generation time) of the cells, whether or not the previous round has been completed. The vertical red lines indicate the times of cell division. The percentages in the key give the extent of chromosome replication.

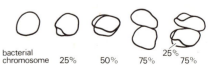

20

minutes and there is a gap of 20 minutes without DNA synthesis before cell division. With cell cycles shorter than 40 minutes new rounds of replication start before an old round is completed. In a 30-minute cycle for instance (Fig. 13(b)), one round of DNA replication begins at the beginning of one cell cycle and is completed 10 minutes after the start of the next cell cycle (20 minutes before cell division). However, as shown in Figure 13(b), a new round of replication can start 10 minutes before completion of the previous round.

In cells with a 60-minute cell cycle DNA synthesis is initiated at the beginning of the cell cycle. This is also true for cells with a 30-minute cell cycle except that initiation occurs at the start of the cycle previous to the one at which the duplicated DNA is segregated, and the cell subsequently divides. The initiation of DNA synthesis is not necessarily associated with the beginning of the cell cycle. For instance, in cells growing with a doubling time of 40 minutes, chromosome duplication is initiated half-way through one cycle and completed half-way through the next cycle (20 minutes before division). (Remember that this model applies only to cells growing with a doubling time faster than 60 minutes.)

This model explains the earlier observation that the DNA content of a cell increases with growth rate. A glance at the chromosome configurations in Figure 13 shows that an analysis of experimental cultures at different growth rates will show increasing DNA per cell at faster growth rates (because of the occurrence of multiple rounds of DNA synthesis in fast-growing cells). One consequence of the fact that in fast-growing cells there may be multiple initiations of DNA synthesis is that there are relatively more copies of genes close to the origin than those close to the terminus of replication. It is interesting that in *E. coli* the genes for ribosomal RNA and proteins are close to the origin of replication. In the absence of other controls such as negative feed-back, these genes are 'enriched' at fast growth rates and so too will be the number of ribosomes per cell. Here we see the interdependency of ribosomal concentration and growth rate, because not only does a fast growth rate appear to be necessary for enhanced ribosome synthesis but also increased numbers of ribosomes are necessary if the faster rate of protein synthesis accompanying fast growth rates is to be achieved.

7.3.2 Chromosome segregation

Once the chromosome is replicated it must be segregated to the two potential daughter cells. In eucaryotes this is achieved by mitosis, but in the absence of mitosis in bacteria another mechanism must be postulated. An early observation gave a clue to how this was achieved—the observation that the chromosome is in some way associated with the cell membrane. It is suggested that at the end of DNA replication the daughter chromosomes are attached side by side on the cell membrane. Cell membrane growth then occurs between the two attachment sites, moving the two daughter chromosomes to opposite ends of the cell. We have mentioned previously that in *Streptococcus* new cell wall is formed equatorially and old material is conserved. Clearly, if the sites of attachment of the daughter chromosomes are at the old conserved parts, equatorial growth will ensure their polar segregation.

7.3.3 Initiation of DNA synthesis

Initiation of DNA replication is dependent on protein synthesis although, once initiated, protein synthesis can be inhibited without affecting DNA replication. Evidence for the role of proteins in initiation is based partly on chloramphenicol, an inhibitor of protein synthesis, which prevents initiation of DNA synthesis but not the completion of a round once it has been initiated. Additional evidence comes from a mutant of *E. coli* that requires the amino acid arginine to be present in the culture medium if protein synthesis is to continue. If this mutant is starved of arginine then new rounds of DNA are not initiated but rounds under way are completed.

It has been suggested that the synthesis of one or more initiation proteins occurs throughout the cell cycle and that DNA synthesis is begun once the initiator has reached a critical level. If initiator is made throughout the cell cycle, under conditions when DNA synthesis is experimentally inhibited but protein synthesis

initiation proteins

is allowed to continue, initiation should be more frequent than normal because of an excess of initiator after removal of the DNA synthesis block. This has been shown to occur using a variety of DNA inhibitors as well as using a strain of *E. coli* which requires thymine for growth; in this instance DNA synthesis is inhibited by starving cells for thymine while protein synthesis continues. The mechanism whereby initiator proteins initiate DNA synthesis is as yet unknown.

7.3.4 Control of cell division

DNA synthesis is a prerequisite for cell division. If DNA synthesis is inhibited before duplication is complete then the cell division which would occur 20 minutes after complete duplication is prevented. Thus completion of a round of DNA synthesis is necessary for cell division.

As cell division occurs 60 minutes after the initiation of DNA synthesis at a wide variety of growth rates, it is tempting to speculate that DNA initiation provides the primary control for division. However, this seems unlikely because there are mutants which can divide when DNA synthesis is inhibited thus providing two normal size cells, only one of which contains DNA. This and other experiments dissociating DNA synthesis and cell division suggest the possibility that there may be a 'division' cycle which is distinct from a 'DNA' cycle. The events of the former cycle are largely unknown but it is known that a constant period of protein synthesis of 40 minutes is necessary if cell division is to take place. The 40-minute period is followed by a 20-minute interval during which further protein synthesis is not required. An exception to this is a short period of protein synthesis at the time of termination of a round of DNA replication. Donachie has suggested that during the 20-minute period the assembly or modification of preformed proteins into some structure necessary for division occurs. The final step in the process requires the participation of the 'termination protein(s)' synthesized 20 minutes earlier at the completion of DNA synthesis.

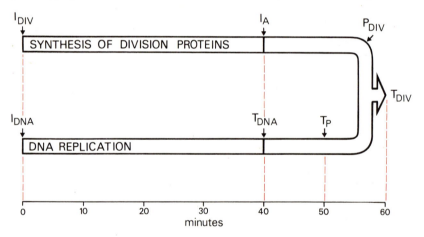

Figure 14 The cell cycle of *Escherichia coli.*

Donachie's model of the *E. coli* cell cycle is shown in Figure 14. At each doubling in mass, two processes are initiated simultaneously. These are the initiation of DNA replication (I_{DNA}) and the initiation of a sequence of events leading to division (I_{DIV}). Termination of chromosome replication (T_{DNA}) at 40 minutes induces the synthesis of termination protein (T_P). The first 40 minutes of the division sequence involve protein synthesis, which is then followed by the initiation of assembly (I_A). After 15–20 minutes more the cell has reached a stage (P_{DIV}) where interaction between some cell wall precursor (or 'septum primordium') and the termination protein leads to cell division (T_{DIV}).

Once every cell cycle, a trigger (T) starts two parallel but separate sequences of events which take constant periods of time to complete, largely independent of the rate of cell growth. One of these processes is chromosome replication and the synthesis of termination protein; and this requires 40–45 minutes to complete. The other is a sequence of protein synthesis followed by another process which may be assembly of some septum precursor. This sequence requires nearly 60 minutes and, at the end of it, there is an interaction between the septum primordium and the termination protein to give the final septum and cell division. This last event takes only a few minutes.

After completing this Section, attempt SAQs 10–13 on p. 32.

Summary of Section 7.3

1 DNA replication at a variety of growth rates takes 40 minutes in *Escherichia coli*.

2 A further period of 20 minutes must elapse before cell division.

3 The growth rate of *E. coli* cells under these conditions may vary from 20–60 minutes.

4 The Cooper–Helmstetter model explains how cells can divide every 20 minutes, and each daughter receive a complete genome. The model requires that in fast-growing cells new rounds of replication are begun before old ones are complete.

5 Chromosome segregation is achieved in the absence of mitosis possibly by growth of membrane between two points at which the replicated bacterial chromosomes attach.

6 Initiation of DNA replication is dependent on protein synthesis, but once initiated replication can go to completion in the absence of protein synthesis.

7 DNA synthesis is a prerequisite for cell division.

7.4 DNA synthesis in eucaryotes

> **Study Comment** This Section reveals differences in DNA synthesis in procaryotes and eucaryotes. In eucaryotes, DNA synthesis always occupies a restricted part of the cell cycle. Information on the initiation of DNA synthesis has come from cell fusion and nuclear transplantation studies. These techniques were described in S2–5, Unit 3.

DNA synthesis is one of the most studied aspects of the cell cycle in eucaryotic cells. It is clear that in both the lower eucaryotes (such as yeast) and the higher eucaryotes (such as mouse cells in tissue culture) DNA synthesis occupies a restricted part of the cell cycle. The DNA synthetic period occurs during interphase and is known as the S phase of the cell cycle. It is preceded by what is known as the first gap, G1, and is followed by the second gap, G2. Mitosis, M, follows the G2 period. Thus the cell cycle is divided into G1, S, G2 and M phases. A variety of methods including biochemical, cytological and autoradiographical have been used to determine the length of each period under different conditions. Mouse L cells have been well studied in this respect and a cell which takes 20 hours to complete a cycle spends roughly 9 hours in G1, 6 hours in S, 4 hours in G2 and 1 hour in mitosis (see Fig. 1 on p. 6). The stage that is most variable in time is G1. In a study of different epithelia of the gut, the cell generation time of the epithelia of the small intestine was 17 hours and of the oesophagus 181 hours. The G1 period of these cells ranged from 7 to 171 hours, but the S and G2 and M periods remained constant at 10 hours. Similarly, if one examines one particular type of cell (for instance Chinese hamster cells in culture) and the growth rate is altered by changing the richness of the growth medium, it is apparent that G1 varies in time while S and G2 and M remain at around 8 hours regardless of the period (between 13 and 24 hours) that the cells take to double.

Our understanding of G1 is poor and few events that uniquely characterize it have been reported. As we have mentioned, G1 lengthens in slowly growing cells and when cell growth is arrested cells normally stop in the G1 phase. It is likely then that G1 contains important events that regulate cell division. In addition, events in G1 probably play an important role in the initiation of DNA synthesis—particularly events towards the end of G1.

DNA is also present in the cytoplasmic organelles of eucaryotes—the mitchochondria and chloroplasts. The difficulty of measuring this DNA has precluded clear evidence of its replication although there are instances of synchronous replication of mitochondrial DNA, for instance in the alga *Chlamydomonas*. In the slime mould *Physarum* mitochondrial DNA synthesis does not occupy a discrete phase of the cycle but occurs throughout the cell cycle. In so much as it has been speculated that these organelles may have originated as procaryotic symbionts it would be interesting to know if they share any features with procaryotic DNA synthesis. Incidentally, the finding that mitochondrial DNA synthesis is continuous throughout the cycle could be because each mitochondrion has a restricted period of DNA synthesis which varies among the mitochondria of a cell.

We will not go into molecular detail of DNA synthesis in eucaryotes; however, it has some features quite different from bacterial DNA synthesis. The replication of individual eucaryotic chromosomes is achieved by a number of replicating units (replicons), that is, DNA replication takes place at a number of different sites. The replicating units are short—there may be 1 000 replicons per chromosome and their rate of DNA replication is an order of magnitude lower than DNA synthesis in bacterial cells.

7.4.1   Initiation of DNA synthesis

How is the initiation of DNA synthesis controlled? Our present knowledge is limited to an understanding of where the initiation signal originates in the cell. The current evidence is that it originates in the cytoplasm. Information has been obtained using two techniques, nuclear transplantation[13] and cell fusion.* In the former, nuclei from one cell are injected into another and the effect of the added nuclei on the recipient cell is observed. Cell-fusion experiments entail the fusion of cells usually of a different type, for instance rapidly dividing and non-dividing cells, or the fusion of cells at different stages of the cell cycle. In these experiments the components of two cells are enclosed within one cell membrane. These techniques were pioneered by Henry Harris and his colleagues in Oxford, who first fused human HeLa and mouse tumour cells.

If nuclei are taken from cells which do not normally divide, such as the neurones and blood cells of the adult frog, and injected into the cytoplasm of frog eggs, the adult nuclei undergo DNA synthesis, presumably under the influence of a cytoplasmic DNA initiator. Similarly, nuclei in which DNA synthesis has stopped (from the brain cells of the South African clawed toad *Xenopus*), if injected into mature unfertilized eggs begin DNA synthesis. When brain nuclei are injected into immature eggs they are not stimulated to divide. Thus during maturation of the oocyte, factors responsible for the initiation of DNA synthesis emerge that are capable of acting on nuclei from adult cells. The cytoplasmic initiation is not species or tissue specific in that mouse liver nuclei can be stimulated after fusion with human HeLa cells. This response can also be elicited *in vitro*—extracts of the cytoplasm of mouse L cells will cause DNA synthesis in hen erythrocyte nuclei.

The role of the cytoplasm in DNA synthesis is also indicated by experiments involving the fusion of cells. If cells which do not undergo DNA synthesis, for example hen erythrocytes or macrophages, are fused with actively growing cells, they are induced to replicate their DNA. Cell-fusion experiments, like nuclear transplants, suggest that the DNA initiation signal is not very specific, as hen erythrocyte nuclei start DNA synthesis when fused with human HeLa cells.

From nuclear transplant experiments with amoebae the following conclusions were made:

1   G2 phase cytoplasm does not contain an inhibitor of DNA synthesis.

2   Once DNA synthesis is completed it cannot be duplicated before cell division.

The experiments performed to reach these conclusions examined DNA synthesis in:

(a)   S-phase nuclei transferred to G2-phase cytoplasm;

(b)   G2-phase nuclei transferred to S-phase cytoplasm;

> QUESTION   Knowing 1 and 2 above, would you expect S-phase nuclei to stop DNA synthesis or to continue DNA synthesis?
>
> ANSWER   S-phase nuclei continue to synthesize DNA.

---

* The use of cell fusion followed the discovery that certain viruses change cell membranes in some way that allows adjacent cells to fuse into one.

QUESTION   Would you expect G2-phase nuclei to start to synthesize DNA or not?

ANSWER   G2-phase nuclei do not start to synthesize DNA.

Similar conclusions result from the work of Rao and Johnson (1970) involving the fusion of human HeLa cells at different stages of the cell cycle. When G2 cells are fused with S-phase cells the G2 nuclei are not stimulated to undergo DNA synthesis, but S-phase nuclei continue DNA synthesis. In addition, when G1-phase cells are fused with cells in S phase the G1 cells initiate DNA synthesis earlier than usual. This suggests that there is a substance present in S-phase cytoplasm that will initiate DNA synthesis in G1 but not G2 nuclei. Once DNA synthesis is initiated, it can be completed in G2 cytoplasm in the absence of this substance. The nature of this substance is at present unknown. We have discussed the possibility that there may be a protein initiator of DNA synthesis in procaryotes and this may be true for eucaryotes as well. According to such a hypothesis, the initiation of replication is dependent on synthesis of an initiator protein. This protein accumulates during the cell cycle and when it is present in sufficient quantity DNA replication is initiated. Experimental evidence for this is based on the use of protein synthesis inhibitors, such as cycloheximide. These inhibitors prevent the initiation of DNA synthesis and depress DNA synthesis that has already been initiated.

In the slime mould *Physarum* if cycloheximide is added at the beginning of the S period about 20 per cent of the DNA replicates. This raises the possibility that there are several rounds of replication during the S period and that protein synthesis is necessary to *initiate* each round but not to complete it. It would be of interest to know if each chromosome required a different protein initiator. Possibly the existence of multiple initiators might explain the asynchrony of DNA replication observed in human chromosomes during the S period.

Autoradiographs of isotopically labelled human chromosomes indicate that there are differences in the labelling pattern during the S period. Parts of chromosomes as well as some entire chromosomes replicate their DNA only late during the S period. This phenomenon is not random—chromosomes that replicate late in one cell cycle replicate late in successive cycles. Not all of the late-replicating regions are heterochromatic*, but heterochromatic regions of chromosomes are late-replicating.

7.4.2   Mitotic signals

Cell-fusion experiments have been used to investigate the events by which a cell enters mitosis. If a cell in mitosis is fused with a cell in the G2 stage, the chromosomes of the G2 nucleus condense much earlier than usual. If a mitotic cell is fused with a G1-phase cell the chromosomes of the G1 nucleus condense but they are seen as single chromosomes because they have not yet replicated. If a cell in mitosis is fused with a S-phase cell then the chromosomes of the latter 'pulverize', they condense in small fragments. Clearly there is something in mitotic cells which forces cells in any phase of the cell cycle to condense their chromosomes. These experiments work even between hybrids, for instance hybrids of toad and man.

Summary of Section 7.4

1   The cell cycle of eucaryotes is divided into G1, S, G2 and M phases. At different growth rates G1 varies in time, but S and G2 and M remain roughly constant.

2   It is likely that the signal for the initiation of DNA replication originates in the cytoplasm. The signal is not species-specific.

3   Cell-fusion and nuclear-transplantation experiments indicate that there is a substance present in S-phase nuclei which will initiate DNA synthesis in G1

---

* HETEROCHROMATIC Parts of chromosomes that stain strongly during interphase are termed heterochromatic.

nuclei. Once DNA synthesis is initiated it can be completed in G2 cytoplasm in the absence of this substance.

4   If cells in mitosis are fused with cells of any other stage of the cycle (G1, S or G2), the chromosomes of the recipient cell condense. Mitotic cells contain something that causes chromosomes to condense.

7.5   What are the mechanisms that control cell division?

Study Comment   This Section discusses a central problem of cell biology. The evidence for proteins involved with the control of cell division is presented in Section 7.5.1. Recent studies on the genetic control of the cell cycle (Section 7.5.2) seek to probe the causal relationships between the events of the cell cycle. These and other studies described in Section 7.5.3 suggest that there may be more than one biological clock[14] within the mitotic cell cycle and that these clocks may, to a certain extent, be independent of each other.

7.5.1   Division proteins

Earlier (Section 7.1) we discussed methods of inducing synchronous division using shocks of one kind or another. A well-studied instance of this is the synchrony that can be induced by repeated heat shocks of the protozoan *Tetrahymena*. Asynchronous cultures growing at the optimum temperature of 29 °C are raised to 34 °C for 30 minutes and then regrown at 29 °C. This process is repeated 8 times at 30-minute intervals. The cells stop dividing and after the series of shocks, synchronous division ensues. This continues for some generations but divisions gradually become less and less synchronous.

Although division is stopped by heat shocks, cell growth is continuous so that upon the first synchronous division the cells are both larger and more variable in size than the original population. However, during the subsequent cell cycles, which are shorter than normal, cell size is reduced to usual dimensions and the variation in cell size is also reduced. The decrease in variability during synchronous growth indicates that the larger cells have a shorter interdivision time than the smaller cells.

Why repeated heat shock causes synchronous growth has been the subject of intense study, and it appears that cells at different stages of the cycle respond differently to heat shock. If a synchronous population is subjected to a heat shock at different stages of the cycle, cells early in the cell cycle show no delay (or very little) in cell division whereas cells later and later in the cell cycle show progressively increased delays before division occurs after a heat shock.

However, there is a point in the cell cycle (the transition point) after which the cells are not delayed by heat shock. This is shown in Figure 15. If cells early in cycle are delayed but a little and cells later in the cycle are delayed more, it is clear that some synchrony is induced by only one heat shock. However, the synchrony will be impaired by the presence of cells which are post-transition point and which will therefore not be delayed. These cells will, however, be early in the cell cycle at the next temperature shock and will be subject to delay so that synchrony is enhanced by repeated shocks. Similar instances of variable delay and transition points have been observed in other cell populations such as human tissue culture cells and Chinese hamster cells, as well as the lower eucaryotes *Paramecium* and the slime mould *Physarum*. Chemical agents, such as inhibitors of protein synthesis, as well as physical means can affect these responses.

Zeuthen (1969) has suggested that in *Tetrahymena* division proteins are synthesized throughout the cell cycle and are vital for cell division. The division proteins are unstable and heat shocks result in complete breakdown. At and after the transition point the proteins are stable. The stability may be a result of the completed assembly of a 'division-protein structure'. This structure then plays some essential role in cell division. This model is complex but there is in *Tetrahymena* an analogous situation of a structure that breaks down if its synthesis is interrupted. This is the 'oral apparatus' which develops at the equator of the cell and forms the mouth of the posterior cell after division. The synthesis of the oral apparatus occurs throughout the cell cycle but if the culture is subjected

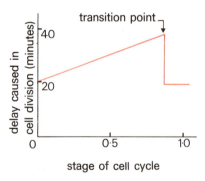

Figure 15   Effect of 20-minute heat shocks at various stages during the cell cycle of *Tetrahymena*. At the beginning of the cell cycle and after the transition point, no delay in excess of the heat-shock period (20 minutes) is observed.

26

to heat shocks the oral apparatus regresses completely. It does, however, have a transition point after which it is insensitive to such shocks. Thus the oral apparatus behaves just like the expected behaviour of division protein. It is unlikely that the oral apparatus contains the division protein because under certain circumstances *Tetrahymena* can divide in the absence of the oral apparatus. Further, the phenomena of delay and transition point are observed in bacteria as well as in human cells that do not possess an oral apparatus.

The identity of these 'division proteins' is a mystery. It is not known whether they are enzyme proteins or structural proteins, or whether they are present in small or large quantities within the cells. A practical problem in the search is that although we know that the biological functions of these proteins are destroyed by heat shock, it may not be because they are degraded or destroyed. Rather it may be because of inactivation without significant change in the component molecules and this will be difficult to detect by the usual physico-chemical protein detection techniques.

7.5.2 Genetic control of the cell cycle

Although many cytological and biochemical markers have been assigned within the eucaryotic cell cycle, the genes that control the progression of events, the nature of their products, and the functions these products perform in the cell cycle remain obscure.

The identification of genes with essential cell-cycle functions requires special methods because most mutations will be lethal and so cannot be isolated. Clearly, if a cell cannot divide we cannot obtain large numbers of its type for biochemical analysis. A convenient approach uses conditional lethal mutants in which gene function is only severely affected under certain conditions, for example, under high temperature. Hartwell *et al.* (1970) used the yeast *Saccharomyces cerevisiae* for a mutational analysis of the cell cycle, selecting temperature-sensitive mutants as a means of examining genes that affect essential cell functions such as cell division.

This temperature-sensitive mutant performs all the cell functions at the normal growth temperature (23 °C): however, at higher temperatures (36 °C) the mutation expresses itself as a defect. The cell cycle of yeast is like that of higher eucaryotes—it has a restricted DNA synthetic period, S, with a G2 interval separating the end of DNA synthesis from nuclear division and a G1 phase separating cytokinesis (cell division) from the initiation of DNA synthesis. Nuclear division in yeast is essentially similar to that in higher eucaryotes entailing the distribution of 17 chromosome pairs to the daughter cells by mitosis. This yeast divides by budding and the bud increases in size throughout the cell cycle until it becomes the size of the mother cell. This provides a visual marker to the position of individual cells in the cell cycle.

Temperature-sensitive mutants were selected after mutagenesis by transferring yeast cultures at 23 °C to pre-warmed agar plates at 36 °C and examining the cells by time-lapse photography. A number of mutants were observed in which all the cells were arrested at a particular stage of the cycle. Hartwell looked for mutants that exhibit a normal cell cycle at the optimum growth temperature but that all arrest at a particular stage of the cell cycle when shifted to the higher (restrictive) temperature. He has, for instance, obtained mutants defective in bud formation when shifted to the restrictive temperature, by selecting cells that accumulate in the single-cell unbudded stage of the cycle.

If one makes synchronous cultures of such mutants in culture at 23 °C and shifts samples to 36 °C at different points of the cell cycle, it is possible to detect a critical stage of the cycle when the temperature-sensitive event occurs at the normal (low) temperature. Thus, cells that have completed this event before the shift to high temperature go through one more cell cycle than do cells which have not completed the event before the temperature shift. In both cases, however, the cells arrest at the same stage of the cycle (see Fig. 16). The stage when cells arrest is known as the 'termination point' and the critical period described above is known as the 'execution point' because it is the time in the cell cycle when the temperature-sensitive function is normally executed. In some mutants the execution point and the termination point occur at the same stage of the

termination point
execution point

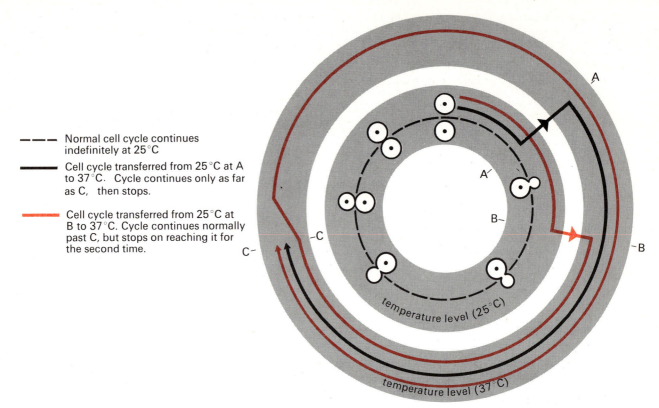

temperature level (25°C)

temperature level (37°C)

cycle but in many cases the latter occurs after the execution point. A gene product is made at the execution point and is necessary for the cell to proceed beyond the termination point.

QUESTION What is the difference between an execution point and a transition point?

ANSWER They are different expressions describing the same phenonenon. Before this point is reached cells are arrested in their progress towards cell division. Once past this, cells cannot be interrupted in their progress towards division by effectors.

Temperature-sensitive mutants have been obtained for genes controlling bud emergence, DNA replication and the initiation of that replication, nuclear division, and cytokinesis. Analysis of these mutants has led to some conclusions about the interdependency of cell cycle events following a block in one of the cell cycle events. If DNA synthesis is blocked, nuclear division cannot occur. Interestingly, blocked cells in which the nuclei are stained reveal an elongated nucleus midway between parent cell and bud, a characteristic of dividing nuclei. Thus it appears that DNA replication is not vital for the initiation of nuclear division but that replicated chromosomes are required for completion. Such cells cannot undergo cell separation but it is likely that this is a direct result of the inhibition of nuclear division. Some mutants defective in DNA synthesis, nevertheless were able to produce buds periodically at the higher temperature. Thus bud emergence is not dependent on DNA replication, nuclear division or cell separation. It is interesting that the periodic appearance of buds at the restrictive temperature matched the normal doubling time, suggesting a cellular clock that controls bud emergence and is independent of DNA synthesis, nuclear division and cell separation. This raises the possibility of more than one biological clock within the cell cycle. The nature of such clocks is still obscure.

Figure 16 The effect of shifting a temperature-sensitive mutant from the low (permissive) temperature to the high (restrictive) temperature at different stages of the cell cycle. At the permissive temperature (the inner circle) cells can go through repeated cell cycles. DNA synthesis occurs about the same time as bud initiation (between A and B) and nuclear division occurs when the bud is almost the size of the parent cell (just after C). The temperature-sensitive cell cycle mutant represented here has an execution point at the time of bud initiation and a termination point at the beginning of nuclear division (at C). If cells are shifted to the higher temperature before the execution point they traverse the cycle until they arrest at the termination point. If the shift occurs after the execution point then the normal gene product necessary for the cell to progress beyond the termination point is made. The cell passes the termination point, divides and arrests in the subsequent cell cycle at C as in the second cell cycle the correct gene product at the execution point is not made. It is interesting that the execution point is early in the cycle. Clearly the preparations for cell division occur early in the cell cycle.

7.5.3 Cell-cycle clocks

During the course of a cell cycle the cells pass through a series of stages or 'markers'. These markers may be morphological, for example, mitosis and in budding yeasts, bud emergence. They may be biochemical, for example, the periodic enzyme synthesis described in Section 7.2.4 or physiological, for example, the transition points discussed in Section 7.5.1. In addition, at least in yeast and the alga *Chlamydomonas*, we have genetic markers defined by the execution and termination points of gene action. An important question which analysis of cell cycle mutants is beginning to answer is: what is the interdependency of markers and what are the causal relationships within the cell cycle?

An extreme example is where there is a direct causal relation between one marker and the next. Here the order of markers during the cell cycle is fixed and a block placed on any of them will prevent the cell's progression through the cell cycle. Analysis of mutants in yeast has indicated that some events, such as nuclear division, are dependent on prior DNA synthesis. However, periodic emergence of buds is not part of this fixed causal sequence as it is independent of DNA synthesis, nuclear division and cell separation. This and other work suggests the possibility that there are different sequences of events that take place during the cycle. It has been suggested by Mitchison (1973) that there may be a DNA-division cycle which is to some extent independent of, or at least can be dissociated from, a growth cycle. Candidates for the DNA-division cycle include DNA synthesis plus nuclear and cell division; in addition, in yeast, we would include some of the gene products whose temperature-sensitive controls have been shown. The growth cycle might include most of the macromolecular synthesis that takes place during the cycle. Periodic enzyme synthesis provides many markers in this cycle.

The biochemical evidence that at least these two cycles exist comes from the work of Mitchison and his colleagues. If DNA synthesis is stopped by the introduction of an inhibitor to a culture, then cells are arrested in their progression through the DNA-division cycle. However, the growth cycle is unaffected and cells continue to grow at their normal rate, thus the two cycles become disassociated. Similar phenomena occur using temperature shocks with *Tetrahymena*. Because all the cells are arrested at DNA synthesis, when the block is removed, cells divide synchronously. However, the growth cycle is not synchronized. On the other hand, if synchronous growth is achieved by selection methods then both cycles are synchronous. Mitchison has suggested that we can use these methods of synchronization to determine whether a marker belongs to one or other cycle. Thus, if it is a marker belonging to the growth cycle it should be synchronous only with selection synchrony and not with induction synchrony, whereas if it is a member of the DNA-division cycle it will be synchronous with both.

Using these techniques it has been shown in fission yeast that three enzymes—alcohol dehydrogenase, ornithine transcarbamylase and aspartate transcarbamylase—are members of the growth cycle. In selection synchrony they are synthesized periodically but in induction synchrony they are synthesized continuously as in a normal exponential asynchronous culture. It is worth noting here that this distinction indicates that in synchronous culture initiated by induction, continuous enzyme synthesis may be observed, but this need not imply that an enzyme of this type is not synthesized periodically during the normal cell cycle. It may be a reflection of the asynchrony of growth cycle events. It is obvious from this brief survey of our knowledge of the controls of cell division that we know very little. Because this is a most fundamental problem in biology it is encouraging that very promising investigations are underway. In particular, it will be a major step forward if some of the biochemical lesions that cause temperature-sensitive blocks in the yeast cell cycle can be identified.

After you have completed Sections 7.4 and 7.5, attempt SAQs 14–18 on pp. 32 and 33.

Summary of Section 7.5

1 *Tetrahymena* can be made synchronous by repeated heat shocks.

2 Analysis of this phenomenon shows that cells early in the cycle show little delay in cell division whereas cells later and later in the cycle show progressively increased delay before division occurs after a heat shock. However, there is a point in the cell cycle (the transition point) after which the cells are not delayed by heat shock.

3 It is suggested that division proteins are synthesized throughout the cycle and are vital to division. The division proteins are unstable and heat shocks result in complete breakdown. At the transition point the proteins become stable.

4   Temperature-sensitive mutants of *Saccharomyces cerevisiae* which arrest at a particular stage in the cell cycle at the restrictive temperature have been observed.

5   Analysis of these mutants has led to conclusions about the interdependency of cell cycle events following a block in one or other of the cell cycle events.

6   It is possible that there is more than one biological clock during the cell cycle—for instance, a growth cycle and a DNA-division cycle. Methods have been devised to see whether an event belongs to one or other of these cycles.

7.6   Conclusions

In this Unit we have dealt mainly with the interphase part of the eucaryotic cell cycle and we have seen that interphase is not a time of uniform growth but that definite changes in the chemical composition of the cell take place. The most obvious example of this is the doubling in DNA content of the cell over a relatively short part of the cycle. In addition, while total protein increases continuously throughout the cell cycle, the individual proteins that comprise total protein are in many instances synthesized periodically, once a cycle. There is then, during interphase, chemical differentiation that precedes the morphological differentiation observed during mitosis. It should be apparent from this that during the cell cycle we find processes of morphogenesis and periodic expression of genes analogous to those observed in whole developing organisms.

Section 7.2.4 dealt in some detail with two theories to account for periodic gene expression. These theories seek to answer the question of why parts of the genome are expressed at one point in time and not another. This is of importance not only to studies of the cell cycle but also during differentiation of other systems where parts of chromosomes remain unexpressed during certain developmental stages but are expressed at other stages.

In discussing enzyme synthesis and DNA synthesis in procaryotes and eucaryotes evidence was presented showing that the synthesis of these cell components may well be very different. With regard to enzyme synthesis it was suggested that compartmentalization of eucaryotic cells, for example, into nucleus and mitochondria, and the presence of large pools of macromolecular precursors effectively buffers the cells from changes in the external environment. So, immediate responses to environmental changes are much less vital in eucaryotes than procaryotes and this may be reflected in the regulation of enzyme synthesis.

Although we know very little about the control of cell division, attempts are being made to elucidate the interdependency of cell cycle events and the causal relationships within the cell cycle. Research on the control of cell division suggests that there may be more than one biological 'clock' during the cell cycle. Indeed, evidence was presented from separate genetical and biochemical studies of yeast that suggests at least two biological 'clocks' or 'cycles' within the mitotic cycle which are to some extent independent of one another: at least they can be dissociated.

Recent research on the genetic control of cell division has identified some of the genes that control the progression of events through the cell division cycle. This is in itself a step forward in our understanding of the cell cycle, but it will be a major step forward if the nature of the gene products and the functions these products perform in the cell cycle can be ascertained.

Self-assessment questions

Section 7.0 (also Unit 1)

SAQ 1 (*Objective 1*) Which of the following structural or functional characteristics are typical of procaryotic cells and which of eucaryotic cells?

(i)   a nuclear membrane
(ii)   mitochondria
(iii)   chromosomes
(iv)   endoplasmic reticulum
(v)   periodic DNA synthesis.

Section 7.1

SAQ 2 (*Objective 3*) Suppose a free living cell increases in volume linearly throughout the cell cycle but dry weight increases as a sigmoidal curve (Fig. 17). Suggest two methods whereby synchronous cultures can be obtained by selection methods.

SAQ 3 (*Objective 3*) Which of the following techniques are examples of induction synchrony methods?

(i)   the addition of an inhibitor blocking DNA synthesis;
(ii)   separation of monolayer tissue culture cells during metaphase;
(iii)   separation of cells on the basis of their size by centrifugation;
(iv)   periodic application of elevated temperature.

SAQ 4 (*Objective 3*) Select from the following list two advantages of using selection methods to obtain synchronous cultures.

Selection methods:

(i)   always give a higher yield of cells than induction methods;
(ii)   are quicker than induction methods;
(iii)   are less likely to produce metabolically abnormal cells;
(iv)   can be applied to a greater variety of cell types;
(v)   always produce populations which display 100 per cent synchrony.

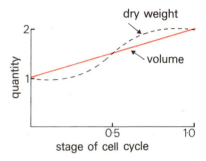

Figure 17 Hypothetical increase in volume (shown in red) and dry weight during a cell cycle.

Section 7.2

SAQ 5 Which of the following cellular constituents (a)–(d) are monitored in the measurement of:

(i)   macromolecular dry mass
(ii)   volume
(iii)   total dry mass?

(a)   low molecular weight pool
(b)   protein
(c)   water
(d)   major macromolecules, e.g. DNA, RNA, carbohydrates.

SAQ 6 (*Objective 5*) Which of the following cell components divide during the cell cycle?

(i)   nucleus
(ii)   mitochondria
(iii)   centrioles
(iv)   chloroplasts.

SAQ 7 (*Objective 5*) In an investigation to study the growth of bacterial cell wall, fluorescent antibody is added to growing cells for a fraction of the cell cycle. The antibody is removed and then the cells are observed using a fluorescence microscope. If cell wall growth is taking place throughout the surface, which part of the cell wall will fluoresce?

SAQ 8 (*Objective 6*) Which of the following statements (i)–(v) relate to:

(a) periodic regulation of enzyme synthesis by the oscillatory repression model;
(b) periodic regulation of enzyme synthesis by the linear transcription model;
(c) neither of these;
(d) both of these.

(i) Production of mRNA is at the site of transcription.
(ii) The level of enzyme production is influenced by inducers or repressors.
(iii) Translocation of part of the chromosome would lead to an alteration in the sequence of timing.
(iv) The position of a gene determines the timing of the enzyme it produces.
(v) There is impressive evidence for its occurrence in bacteria.

SAQ 9 (*Objective 6*) It has been suggested that periodic synthesis of enzymes is a result of oscillations in inducers or repressors. In a bacterial mutant which produces an enzyme constitutively, what would be the expected pattern of enzyme accumulation during the cell cycle—continuous or periodic?

Section 7.3

SAQ 10 (*Objective 7*) Which of the following statements referring to the Cooper–Helmstetter model of DNA replication in bacteria are correct?

(i) A replication fork takes 40 minutes to travel from the origin to the terminus.
(ii) With a doubling time of 60 minutes, there is no DNA synthesis until the end of the cycle.
(iii) With a doubling time of 40 minutes, the rate of DNA synthesis doubles in mid-cycle.
(iv) There is a constant period of 20 minutes between the end of a round of replication and the following cell division.
(v) The model explains why there is an increasing amount of DNA per cell at faster growth rates.

SAQ 11 (*Objective 11*) Which of the following processes (i)–(iv) occur in *E. coli*:

(a) after 40 minutes from the initiation of DNA replication;
(b) 20 minutes before cell division;
(c) for more than 30 minutes;
(d) 60 minutes before division?

(i) the initiation of events leading to cell division;
(ii) the termination of chromosome replication;
(iii) the synthesis of division proteins;
(iv) the initiation of assembly.

SAQ 12 (*Objective 7*) *E. coli* B/r is growing with a generation time of 40 minutes. Before cell division, chromosome segregation occurs. When was DNA synthesis initiated in these chromosomes?

SAQ 13 (*Objective 7*) In *E. coli* B/r, growing with a generation time of 30 minutes, the number of replication forks observed depends upon the particular point of the cell cycle at which the cells are examined. What is the maximum and minimum number that will be observed?

Sections 7.4 and 7.5

SAQ 14 (*Objective 8*) Which of the following features (i)–(iii) relate to DNA synthesis in eucaryotes, which relate to DNA synthesis in procaryotes and which relate to both?

(i) Protein synthesis is necessary for the initiation of DNA synthesis.
(ii) Replication of the chromosome is achieved by a number of replicating units.
(iii) DNA per cell increases at faster growth rates.

SAQ 15 (*Objective 2*) In the accompanying diagram (Fig. 18), identify the phases G1, G2 and S, and measure their approximate duration.

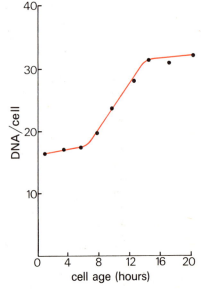

Figure 18 Change in DNA levels during the interphase of mouse L-929 cells.

32

SAQ 16 (*Objective 9*) Suppose there is a substance present in S-phase cytoplasm that will initiate DNA synthesis only in G1-phase nuclei. Predict the results of fusing (a) a G1 cell with an S cell and (b) a G2 cell with an S cell.

SAQ 17 (*Objective 12*) If a cell is shifted from 23 °C to 36 °C before it is halfway through the cell cycle (i.e. before 0.5 of the cycle), it arrests at 0.9 of the cycle and does not divide. If it has passed 0.5 of a cycle before the shift to 36 °C it will divide and arrest at 0.9 of the next cycle. Use this information to decide if (i)–(v) are true or false.

  (i)   The mutant is a conditional lethal mutant.
  (ii) The mutant is a temperature-sensitive mutant.
 (iii) The execution point is at 0.5 of the cell cycle.
 (iv) A critical gene for cell division is expressed at 0.5.
  (v) Another execution point occurs at 0.9.

SAQ 18 (*Objective 13*) Look at Figure 19 and examine its caption. Do the following descriptions (i)–(viii) relate to period A, B, or C or to none of them?

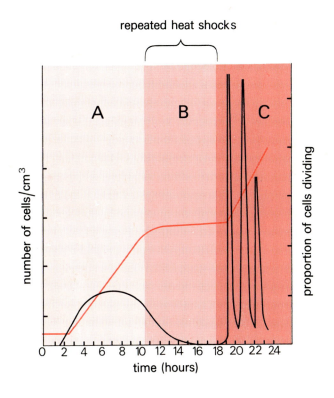

Figure 19 Synchrony of *Tetrahymena* induced by repetitive heat shocks (29–34 °C). The black line shows the proportion of cells in the population that are dividing; the red line shows cell number.

   (i)   a period of repeated heat shocks
  (ii) a period when there is no increase in cell mass
 (iii) a period of gradual decrease in cell synchrony
 (iv) a period when fewer and fewer cells avoid having their division delayed
  (v) a period when all cells have passed the transition point
 (vi) a period when cells gradually stop dividing
(vii) a period when the time between successive synchronous divisions gets greater and greater
(viii) a period when cells are comparatively large and variable in size.

SAQ 19 Mark each of the following statements as true or false:

(a)   Labelled precursors of DNA are incorporated into DNA only during the S phase.
(b)   Yeasts have some of the cellular features of higher eucaryotes, for example, mitochondria, a nuclear membrane and endoplasmic reticulum.
(c)   Total cell protein increases continuously throughout the cell cycle.
(d)   RNA synthesis in higher eucaryotes occurs only during metaphase when the chromosomes are highly condensed.
(e)   Increase in nuclear volume and dry mass appears to be restricted to the second half of the cell cycle.
(f)   The nuclear membrane of yeasts breaks down just before mitosis.

(g) If periodic enzyme synthesis follows the oscillatory repression model, the addition of inducer should lead to enzyme synthesis.

(h) Fluctuations in an end-product pool always result in periodic enzyme synthesis.

(i) Enzymes that are produced constitutively cannot be synthesized periodically according to the oscillatory repression model.

(j) The linear transcription model predicts that there should be one period of enzyme synthesis per structural gene per chromosome.

(k) According to the linear transcription model, genes on the same chromosome that have a high recombination frequency should be synthesized almost simultaneously.

(l) The DNA content of eucaryotic cells is doubled during a discrete, restricted phase of the cycle.

(m) About one-quarter of the eucaryotic cell cycle is spent in mitosis.

(n) The main difference between fast-growing and slow-growing cells is the length of the G1 period, which can vary from zero to many hours.

(o) Hen erythrocyte nuclei are stimulated to synthesize DNA after fusion with human HeLa cells.

(p) Division proteins in *Tetrahymena* are known to be part of the oral apparatus.

(q) Before the transition point, cell division in *Tetrahymena* cannot be delayed by repeated heat shocks.

(r) Unlike procaryotes, it is known that there is no protein initiator of DNA synthesis in eucaryotes.

Answers to SAQs

SAQ 1
(i) Eucaryote.
(ii) Eucaryote.
(iii) Both. The chromosome of procaryotes contains only 'naked' DNA but that of eucaryotes contains DNA associated with proteins such as histones and acidic proteins.
(iv) Eucaryote.
(v) Eucaryote. However in *Escherichia coli* B/r, growing with a doubling time longer than 40 minutes, there is a period at the end of the cycle without DNA synthesis so that in this case DNA synthesis is periodic.

SAQ 2
(i) As volume changes throughout the cell cycle, cells can be separated according to size (i.e. age) during the cell cycle by centrifugation (see p. 8).
(ii) The pattern of increase of volume and dry weight indicates that the density of cells is changing during the cycle. Cells of a particular age in the cycle can therefore be separated by density. After centrifugation to equilibrium in a density gradient, cells of lightest density can be removed and used to inoculate a synchronous culture (see p. 13).

SAQ 3
(i) Induction synchrony.
(ii) Selection synchrony.
(iii) Selection synchrony.
(iv) Induction synchrony.

If your answer is incorrect, study Table 1 on p. 10.

SAQ 4 (iii) and (ii) are advantages. Selection methods are usually quicker as induction methods require that cells progress around the cell cycle until they reach the block. In animal cells this can be 20 hours.
(i) is false—induction methods give the higher yield (p. 9).
(iv) Both methods can be applied to a variety of cells. For instance, size separation and DNA inhibition can be used for a variety of cell types (Table 1).
(v) Neither induction nor selection synchrony methods produce populations displaying 100 per cent synchrony (see Fig. 5 on p. 11).

SAQ 5
(i) (b), (d).
(ii) (a), (b), (c), (d).
(iii) (a), (b) and (d).
See Fig. 6 on p. 12.

SAQ 6 (i), (iii) and (iv) divide (see Section 7.2.2). However, a new centriole is formed *de novo* near but not touching, one end of the old centriole.

SAQ 7 All parts will fluoresce if cell-wall growth is taking place throughout the surface (see p. 14).

SAQ 8
(i) relates to both of these (d).
(ii) relates to the oscillatory repression model (a).
(iii) relates to the linear transcription model (b).
(iv) relates to the linear transcription model (b).
(v) relates to the oscillatory repression model (a).

If your answers are incorrect, read through Section 7.2.4.

SAQ 9 Continuous. An enzyme that is produced constitutively should be unaffected by possible fluctuations in the level of inducers or repressions (see p. 16).

SAQ 10
(i) Correct.
(ii) Incorrect. DNA synthesis occurs for the first 40 minutes. Cell division occurs after a further 20 minutes.
(iii) Correct. The rate doubles in mid-cycle because at mid-cycle, at the completion of one round of DNA synthesis, initiation of DNA replication occurs in the newly replicated molecules.
(iv) Correct.
(v) Correct. At faster growth rates new rounds of DNA replication begin before old rounds are complete but at slow growth rates, for example 60 minutes, there is only one round of replication per cycle. This results in more DNA per cell at faster growth rates (see Fig. 13 on p. 20).

If your answers are incorrect, read through Section 7.3.1.

SAQ 11
(i) (d) occurs 60 minutes before cell division as it coincides with the initiation of DNA synthesis.
(ii) (b) occurs 20 minutes before cell division.
(iii) (c) occurs for more than 30 minutes.
(iv) (a) occurs after 40 minutes from the initiation of DNA replication.

If your answers are incorrect, read through Section 7.3.4.

SAQ 12 DNA synthesis was initiated in mid-cycle in the previous cycle. It takes 40 minutes, so synthesis terminates in mid-cycle and cell division occurs 20 minutes later (Fig. 13).

SAQ 13 The maximum is two because a new round begins before an old round is complete. The minimum is one—there is a 20 minute period in which rounds of DNA replication do not overlap (Fig. 13).

SAQ 14
(i) Protein synthesis is implicated in the initiation of DNA synthesis in both procaryotes and eucaryotes although the evidence is better for procaryotes.
(ii) Only in eucaryotes is chromosome replication achieved via a number of replication forks. In procaryotes replication begins at the origin and proceeds bidirectionally until the pair of replication forks reach the terminus (see Fig. 13 on p. 20).
(iii) This is true of procaryotes and is implicit in the Helm-stetter–Cooper model. In eucaryotes DNA synthesis is always periodic and the amount of DNA per cell does not vary at different growth rates.

SAQ 15 G1 is roughly 8 hours. The phase of increasing DNA content is S, which lasts about 6 hours. G2 lasts about 5 hours.

SAQ 16
(a) The nuclei of the G1 cell will initiate DNA synthesis.
(b) The G2 nuclei will not initiate DNA synthesis. If S-phase nuclei continue to synthesize DNA it indicates that G2 cells do not contain an inhibitor of DNA synthesis.

SAQ 17 (i)–(iv) are true; (v) is false. See p. 27.

SAQ 18
(i) Periodic heat shocks occur during B.
(ii) Neither A, B or C. Heat shocks inhibit division but not growth in mass at any time.
(iii) C. The synchronous divisions initiated at the beginning of the C become less synchronous throughout C.
(iv) This occurs during B.
(v) Neither A, B, or C.
(vi) Cells stop dividing during B because repeated heat shocks produce a delay in division which fewer and fewer cells escape.

(vii) This refers to C. (Successive divisions occur rather more rapidly than the mass doubling time at the beginning of C. It is as if the cells are trying to restore the mass per cell ratio of the A period. During B, growth continues but not cell number so that the mass per cell increases.)

(viii) B. Cells at the beginning of B are at all stages of the cycle and their division is inhibited throughout B but growth is not. At the end of B, cells are larger than during A and more variable in size.

SAQ 19

(a) True, see p. 27.

(b) True, see p. 6.

(c) True, see p. 6.

(d) False, see p. 13. RNA synthesis occurs at all times *except* metaphase.

(e) True, see p. 13.

(f) False, see p. 13. The nuclear membrane in yeasts remains intact throughout mitosis.

(g) True, see p. 16.

(h) False, see p. 16.

(i) True, see p. 16.

(j) True, see p. 17.

(k) False, see p. 17. According to the model, genes transcribed at about the same time are likely to be closely adjacent on the chromosome.

(l) True, see p. 23. This is the S phase.

(m) False, see p. 6.

(n) True, see p. 23.

(o) True, see p. 24.

(p) False, see p. 26.

(q) False, see p. 26. After the transition point is attained, cell division cannot be delayed.

(r) False, see p. 24.

Bibliography and references

1 General

An excellent comprehensive book on the cell cycle is: Mitchison, J. M. (1971) *The Biology of the Cell Cycle*, Cambridge University Press.

A recent review we found interesting is: Mazia, D. (1974) 'The cell cycle', *Sci. Amer.* (Jan. 1974).

2 References cited in the text

Cooper, S. and Helmstetter, C. E. (1968) 'Chromosome replication and the division cycle of *Escherichia coli* B/r', *J. Mol. Biol.*, **31**, pp. 519–40.

Cox, G. C. and Gilbert, J. B. (1970) 'Non-identical times of gene expression in two strains of *Saccharomyces cerevisiae* with mapping differences', *Biochem. Biophys. Res. Commun.* **38**, pp. 750–7.

Donachie, W. D., Jones N. C. and Teather, R. T. (1973) 'The bacterial cell cycle', in *The Society of General Microbiology Symposium 23, Microbial Differentiation* (ed. Ashworth, J. M. and Smith, J. E.), pp. 9–44. Cambridge University Press.

Hartwell, L. H., Culotti, J. and Read, B. J. (1970) 'Genetic control of the cell cycle in yeast, 1 Detection of mutants', *Proc. Natl. Acad. Sci.* U.S.A., **66**, pp. 352–9.

Howard, A, and Pelc, S. R. (1953) 'Synthesis of DNA in normal and irradiated cells and its relationship to chromosome breakage', *Heredity, Lond.* (Suppl.) **6**, pp. 261–73.

Luck, D. J. L. (1963) 'Genesis of mitochondria in *Neurospora crassa*', *Proc. Natl. Acad. Sci. U.S.A.* **49**, pp. 233–40.

Mitchison, J. M. and Vincent, W. S. (1965) 'Preparation of synchronous cell cultures by sedimentation', *Nature, Lond.* **205**, pp. 987–9.

Mitchison, J. M. (1973) 'Differentiation in the cell cycle', in *The Cell Cycle in Development and Differentiation* (ed. Psalls, M. and Billett, F. S.), pp. 1–13. Cambridge University Press.

Rao, P. T. and Johnston, R. T. (1970) 'Mammalian cell fusion: studies on the regulation of DNA synthesis and mitosis', *Nature, Lond.* **225**, pp. 159–64.

Zeuthen, E. and Williams, N. E. (1969) 'Division-limiting morphogenetic processes in *Tetrahymena*', in *Nucleic Acid Metabolism, Cell differentiation and Cancer growth* (ed. Cowdry, E. V. and Seno, S.), pp. 203–16. Pergamon Press.

3 References to material in prerequisite science courses

1 S2–5, TV programme 1
2 S100, Appendix 2
3 S2–5, Home experiment
4 S100, Unit 17, Section 17.11
5 S100, Unit 17, Section 17.11
6 S100, Unit 17
7 S100, Unit 17, Section 17.11
8 S22–, Unit 4, Section 4.7.1
9 S2–5, Unit 5, Figure 30
10 S2–5, Unit 2, Section 2.4.2
11 S100, Unit 19, Section 19.2.3
12 S2–5, Unit 3, Section 3.3
13 S2–5, Unit 2, Section 2.2.1
14 S22–, Unit 8, Section 8.3

Acknowledgements

Grateful acknowledgement is made to the following sources for material used in this Unit:

Figures 6, 7, 8 and 18 Cambridge University Press for J. M. Mitchison, *The Biology of the Cell Cycle*; *Figures 9, 11 and 12* Academic Press Inc. (London) Limited and the authors for Halvorson, Carter and Tauro in *Advance in Microbial Physiology*, **6,** 47; *Figure 13* Academic Press Inc. (London) Limited and the authors for Cooper and Helmstetter in *J. Mol. Biol.*, **31,** 519; *Figure 14* Cambridge University Press for Donachie, Jones, *et al.* in Ashworth and Smith (eds), *Microbial Differentiation*; *Figure 19* John Wiley and Sons, Inc. for E. Zeuthen (ed), *Synchrony in Cell Division*.

S321 Physiology of Cells and Organisms